AMERICAN REPRESENTATION IN OCCUPIED GERMANY

1920 - 1921

VOLUME I

COMPILED BY THE ASSISTANT CHIEF OF STAFF,G-2

AMERICAN FORCES IN GERMANY

AMERICAN REPRESENTATION

IN

OCCUPIED GERMANY.

1920 - 1921.

PREFACE.

When, in December 1918, the Allied and Associated Armies of Occupation marched into Germany and occupied the Rhineland under the terms of the Armistice Agreement, the United States was represented by the Third American Army. On July 3, 1919, that army as a unit having ceased to exist, the American Forces in Germany came into being and took over the duties of occupation.

The Treaty of Versailles had been signed on June 28. Annexed to it was the Rhineland Agreement which provided in part that "a civilian body styled the Interallied Rhineland High Commission, and hereinafter called the High Commission, shall be the supreme representative of the Allied and Associated Powers within the occupied territory. It shall consist of four members representing Belgium, France, Great Britain and the United States." The Treaty came into force on January 10, 1920, and on that day the period of military government in the portions of the Rhineland occupied by the forces of Belgium, France and Great Britain came officially to an end. As the United States had failed to ratify the Treaty, an anomalous situation in the American occupied territory was created. However, in accordance with a modus vivendi proposed by General Allen, the commander of the American forces, and agreed to with appreciation by the High Commission, the ordinances of that body, with certain exceptions and modifications, were permitted to become effective in the American area as well as in other portions of the Rhineland.

The period from the beginning of occupation to the coming into force of the Rhineland Agreement has been ably and exhaustively described in a work of four volumes entitled "American Military Government of Occupied Germany, 1918-1920", which is on file at the American Headquarters at Coblenz and in the War Department at Washington. This valuable account of the American participation in the occupation of the Rhineland was prepared under the supervision of Colonel Irvin L. Hunt, U.S. Army, who served as Officer in Charge of Civil Affairs from the beginning of the occupation until March 1920.

The period of military government in the Rhineland was officially ended on January 10, 1920, but due to the peculiar conditions of the American participation in the occupation, in some particulars, it continued to exist in the American area. Therefore the American control over the German authorities was a mixture of civilian and military control. With the progress of time, more and more powers were transferred from the American military authorities to the American Department of the High Commission, but it proved impracticable to transfer them in their entirety.

The present work of two volumes covers the period from January 10, 1920, to the end of 1921, and so is a sequel to the "American Military Government of Occupied Germany, 1918-1920." As will be seen, its format and methods of description are different. Each account complements the other, and both should be read by those who would obtain a fair idea of a unique phase of the American intervention in European affairs.

Chapters 1 and 2 are contained in Volume I. Chapter 1 is devoted to a narrative of the important events, and of political and economic conditions as well, in both occupied and unoccupied Germany, during 1920 and 1921. Chapter 2 deals with the interior organization, activities and salient policies of the Interallied Rhineland High Commission. There are a number of appendices to this chapter, some of which, because of their length, are contained in separate and smaller volumes.

Chapters 3 to 28 will be found in Volume II. Chapters 3 to 15 describe those activities of the Americans that had connection in one way or another with the German authorities or the population of the area.

The previous work contains but little with regard to the American troops, their problems or their activities. It would seem that for a complete account of the American occupation, a description of the American forces and their activities is an important requisite, and in that belief Chapters 16 to 28 were compiled. In general only those activities peculiar to an army of occupation are described, and for obvious reasons it was necessary in many cases to touch upon the period covered in the previous work.

The officers and civilians who were prominently connected with the American participation in the occupation of the Rhineland are shown at the ends of the chapters dealing with the activities that came within their particular purview.

Inasmuch as these two volumes deal with practically every phase of the American participation during the period covered, the work has been given the title of "American Representation in Occupied Germany, 1920-1921." Chapter 1 was written, and the other chapters compiled, by Major Philip H. Rugby, Infantry, Assistant Chief of Staff, G-2, of the American Forces in Germany.

The following abbreviations are used:

A. C. of S. - - - - - - - - -Assistant Chief of Staff.

A. E. F. - - - - - - - - -American Expeditionary Forces.

A. F. G. - - - - - - - - -American Forces in Germany.

A. M. G. - - - - - - - - -"American Military Government of Occupied Germany, 1918-1920".

D. C. I. - - - - - - - - -Division of Criminal Investigation.

G-1, G-2, etc. - - - - - - Divisions of the General Staff.
G-1, Administration and Personnel; G-2, Military Intelligence; G-3, Operations and Training; G-4, Supply.

O. C. C. A. - - - - - - - - -Officer in Charge of Civil Affairs.

P. M. D. - - - - - - - - -Provost Marshal's Department.

COBLENZ

May, 1922.

TABLE OF CONTENTS

AMERICAN REPRESENTATION IN OCCUPIED GERMANY

1920 - 1921.

VOLUME I.

VOLUME II.

CHAPTER NO. 1

NARRATIVE.

NARRATIVE.

Ratification of the Treaty of Versailles.

On January 9, 1920 the conditions of occupation in the Rhineland were technically the same as they had been during the preceding year. With the signing of the Treaty of Peace and its ratification by the German Government, the severity of the measures adopted for the security of the occupying forces had been relaxed in increasing measure and the restrictions upon the daily life and the movements of the German inhabitants had been greatly lessened. Nevertheless the technical status was that of an armistice between warring nations, the various occupied zones remaining under military government and the French Commander-in-Chief of the Allied Forces of Occupation continuing to be technically the supreme authority within occupied Germany. Actually the powers of the Inter-Allied Rhineland Commission (not to be confused with the Inter-Allied Rhineland High Commission) and other inter-allied commissions had grown greater as time progressed, while in various respects the powers of the military authorities had correspondingly decreased. The commanders of the four occupying forces, notably the American and British, had displayed an increasing tendency to conduct the military government of their respective zones in accordance either with their own or their national policies, rather than in strict compliance with the policies or desires of the French Commander-in-Chief. Divergent views and policies had already developed, but the only effects perceptible beyond officials circles were an impression that the Allies differed in questions of policy and an increased hope thereby given those Germans who believed that the clash of different policies might result in rifts in the Entente through which Germany might derive ultimate good. In general, however, and largely due to the initiative and efforts of the American Commander, cordial and harmonious relations prevailed among the four armies of occupation.

Such was the situation when on January 10, 1920, the exchange of the ratifications of the Treaty of Versailles brought about peace between Germany and the Allied powers, particularly Great Britain, France and Belgium. Automatically the Rhineland Agreement came into force and the reins of government of the occupied territories passed from the hands of the military authorities to those of the Inter-Allied Rhineland High Commission. The anomalous situation in the American zone, caused by the failure of the United States Senate to ratify the Treaty of Versailles, was transformed, by the action of the American Commanding General (Major General H.T. Allen) in publishing the ordinances of the High Commission over his own signature, into one wherein for all practical purposes the conditions were the same as those which prevailed in the other occupied areas. In the final analysis, however, the American forces continued to occupy German territory under the Armistice Agreement of November 11, 1918, as amended by subsequent agreements made between the Allied and German Armistice Commissioners. The actual powers of General Allen continued to be those of the commander of a military force in occupation of enemy territory during an armistice, which means that he remained the supreme authority within the American zone. Under a modus vivendi, proposed by him, approved by the United States War Department and accepted by the High Commission and the Supreme Council, conditions in the American area became practically the same as in the other zones, but technically the supreme power remained in his hands and could be exercised whenever his judgment dictated the advisability or necessity of such a course of action. These are facts that should be kept constantly in mind by the reader or student of the history of the American Occupation subsequent to January 10, 1920.

The coming into force of the ordinances of the High Commission immediately removed two important restrictions upon the German population - censorship and regulations relative to civilian circulation. These were abolished without delay and in those respects conditions in the American area became as similar to those which prevail in time of peace as is possible in an occupied territory. During the period between the signing of the Treaty of Versailles and its ratification by the Allies, the restrictions regarding censorship and civilian circulation in the American zone had gradually become less stringent than in the other occupied areas; hence the changes which took place there in January of 1920 were not so great as in the other zones. There was no change, however, in the restrictions upon the entry into the occupied territories of menbers of the German military and naval forces, who were still required to secure permission therefor from the military authorities. But, very soon thereafter the limit upon the number of such permissions that could be granted within a given period of time was removed by the American military authorities, an action that was gratefully appreciated by many German soldiers and sailors who for several years had been unable to visit their relatives.

During the period of the Armistice, Allied and neutral nationals desiring to enter the occupied territories had been required to secure the permission of the military forces. These restrictions had been gradually relaxed, but not until the coming into power of the High Commission were they entirely abolished, the only remaining requirement being that Allied nationals carry safe conducts or their national passports, while others, whether formerly neutral or hostile, were required to have their national passports. Germans (except those coming from other countries, in which case they were required to have German passports) could circulate freely if provided with identity cards. No visas of any kind were required. It may be remarked in passing that all control posts (except those for the examination of goods) in the American area were removed, while certain ones in the other zones were continued in operation.

One result of the removal of the restrictions upon circulation was that numbers of American business men passed through the American area en route to unoccupied Germany, while as time went on an increasing number of American tourists came to view the unique spectacle of American troops in occupation of German territory. On the other hand many Germans approached the military authorities with a view to securing permission to go to the United States. They were referred to the American Commission in Berlin with the explanation that the granting of permission to enter the United States was not a function of the War Department.

Summing up, it may be said that while a technical state of war, as affected by an indefinite armistice, between the United States and Germany continued to exist, the conditions in the American zone were practically the same as those which prevailed in the other areas, although these were occupied by the military forces of countries technically at peace with Germany. As a matter of fact, conditions in the American and British areas were less burdensome and restrictive upon the German inhabitants than those in the French and Belgian zones.

The First Quarter of 1920.

The first three months of 1920 constituted a period of

transition in the American area. A wise intention to prevent too abrupt a change in the transfer of governing power from the military authorities to the High Commission, as well as the force of circumstances (such as the lack of sufficient personnel in the American Department of the High Commission), caused a very gradual handing over to that Department of the duties which had theretofore appertained to the Officer in Charge of Civil Affairs. At the end of March the Functions of Civil Affairs remained largely the same as before January 10.[1]

The American area suffered much damage from inundations during January, when the waters of the Rhine and Mosel rivers reached the highest level in 136 years, the Rhine registering 9.24 meters (approximately 30 feet) above normal. Not yet recovered from the December floods, the farmlands and the cities and towns along the two rivers were damaged even more than during the preceding month. Industries were forced to suspend, while the shortage of food and fuel was rendered more acute by reason of the greatly increased difficulties of transportation. Heavy losses in the potato crop occured, and in many cases the stocks stored in warehouses and cellars were ruined. By strenuous efforts the problem of the transportation of coal was solved, and the receipt of potatoes from unoccupied Germany and from Holland partly alleviated the acute food shortage. The Central Government as well as the local governments made large appropriations for the relief of suffering

[1] For a more detailed description of the coming into power of the Inter-allied Rhineland High Commission, see Volume I, Chapters 1 and 2; Volume II, Chapter 3; and Volume VII. and A.M.G., Volumes I (Chapter 18) and IV (Appendix 51).

caused by the inundations, and by the end of the month the American area was rapidly recovering from the effects of the great floods.

The service of supply of the American forces was greatly hampered by the two floods, warehouses and accumulations of food and other supplies being much damaged. A great many drums of gasoline floated down the Rhine, but a number of these were recovered later. [1]

The approach of peace had engendered much discussion among the German inhabitants. At first they thought that the new conditions would be almost like those of peace time, and a bolder attitude towards the American troops soon became apparent. The publication of the proclamation and the first ordinances of the High Commission gave rise to much indignant comment by the Germans, who claimed that they would make the conditions of life in the occupied areas far too severe. However, the abolition of censorship of the mail, the telegraph and the telephone and the lifting of the restrictions upon the movements of individuals served to allay their fears in large measure, even if to the ordinary observer there was no apparent change of conditions in the American zone. Further reflection and study of the ordinances of the High Commission showed the Germans throughout the occupied territory that under the new regime officials and private citizens would be less restricted in their duties and affairs than had been the case under military government. Much of the unfavorable

[1] Volume II, Chapter 14.

comment had been due to propaganda; thereafter, except in a few special cases, unfriendly comment anent the ordinances or actions of the High Commission was as much inspired by propaganda as by legitimate cause for complaint.

The failure of the United States Government to ratify the Treaty of Peace which had been signed by its representatives at Versailles, rather mystified the inhabitants of the American area. There soon began to be bruited abroad rumors that the American troops were to be withdrawn and that French troops would replace them. These rumors became periodically recurrent, and each fresh one that arose served to cause uneasiness among the Germans who invariably asserted that they desired American troops to remain in the Coblenz area as long as the troops of any nation remained in occupation of the Rhineland.

While the conditions of life were far more peaceful and the vital problems of employment, food, clothing and shelter were generally less difficult of solution in the occupied territories than in unoccupied Germany, yet existence for the German inhabitants was not easy. Food was costly and not plentiful. There was a constant shortage of fuel, causing the house-holder's allotment to be much below his normal needs, while industries not only had to operate below capacity but also frequently had entirely to suspend operations until their exhausted stocks of coal could be replenished. Prices in general were high and, subject to some fluctuations, grew steadily higher. As has always been the case whereever Americans have predominated in a foreign community, prices in the American area were higher than in the other zones where, in turn, the cost of living was generally higher than in unoccupied Germany. On the other hand, life in the Rhineland was entirely serene as compared with that in the strife-torn remainder of Germany, and the measures taken by the military authorities and the High Commission caused a practically constant operation of public utilities. Food was brought in by the occupying forces and sold through the German administration; when the American forces, because of the statement by the German officials that their Government could no longer afford the high cost of American food, had ceased this practice, large shipments of food were received from the United States and placed upon the market. These measures, combined with the food received through the efforts of the German Government from unoccupied Germany and from Holland, all served to make the lot of the Rhinelander considerably easier than that of his brother who was free from the burden of occupation.

Unemployment was never a serious problem in the American area. Despite their many difficulties, particularly the lack of sufficient coal and of raw material, the various industries displayed a remarkable vitality and kept in operation most of the time.

of sufficient coal and of raw material, the various industries displayed a remarkable vitality and kept in operation most of the time.

The ever-increasing cost of living caused an almost continual succession of wage controversies, with the usual threats of strike. At this time there was a strike fever throughout Germany, but in the Rhineland, and more particularly in the American Area, the wage controversies were usually adjusted without resort to the strike. The American authorities were frequently requested to use their good offices in the way of mediation; whenever they did so their efforts almost invariable met with success. There was comparatively little radical agitation in the American zone, and with very few exceptions every strike or threat to strike was caused by a legitimate wage controversy which had no political tinge.

During this period a number of German prisoners of war, who had been released by the French following the ratification of the Treaty of Versailles, returned to their homes in the American area. While it had been feared that their return might cause some little excitement among the inhabitants, praticularly as the former prisoners were rumored to have become infected with Bolshevism, no Untoward event happened and their return caused hardly a ripple in the prevailing calm.[1]

It will be remembered that the prisoners of war held in France by the Americans had been released during the late summer of 1919.

These three months were marked, among other respects, by violent fluctuations in the exchange rate of the German mark, the pre-war value of which was approximately 25 cents or 4 to the dollar. From 50 to the dollar on January 2, the mark had dropped to 109 on February 7. By March 11 it had risen to 68 to the dollar, falling to over 100 during the political disturbances that commenced on March 13, but rising again to 70 by March 31. These great fluctuations detrimentally affected trade and industry, and consequently the cost of living which rose steadily during the quarter.

The separatist agitation [1] was continued throughout the Rhineland, but it met with little success. Meetings were held and newspapers established, threats of a coup d'etat were again made, and it was said that paper money for the Rhine Republic had even been printed and made ready, but the net result of the agitation and propaganda was nothing more than the gain of a few adherents who were politically of no real prominence. Due to the ill-concealed support of the French and the political blunders of Dorten and his supporters and backers, what may be called the ligitimate separatist sentiment had been replaced in the hearts of the great majority of Rhinelanders by a feeling that first of all they were Germans and loyal to the Fatherland. The average Rhinelander desired to remain within the German Reich, and he felt that separation from Prussia could well await that far day when the foreign forces of occupation had been withdrawn within the confines of their own countries.

[1] A.M.G., Volume I, Chapter 15.

Generally speaking, the first quarter of 1920 passed quietly in the American zone. The political disturbances in unoccupied Germany in March caused some excitement in the Coblenz area, but the reaction to the events beyond the Rhine was only slight. The thrifty, industrious and well-disciplined Rhinelander, resting in the security afforded him by the presence of the occupying forces, was well content to observe events from afar and, without taking any active part himself, calmly to await their final outcome.

Official visits were frequently exchanged between the American Commander and the respective Commanders of the other armies of occupation. Thus in January he and his staff visited General Degoutte, commanding the Allied Forces of Occupation as well as the French army. On March 9 Marshal Foch paid an official visit to General Allen, being followed the next day by Field Marshal Sir Henry Wilson, Chief of the Imperial General Staff of Great Britain. Other distinguished visitors during the period were Generals Leorat and Bray of the French Army and Mr. Frederick Palmer, the noted American war correspondent and author. On March 17 General Allen made a hurried trip to Paris, returning to Coblenz on the 19th, having been called by the American Ambassador for a conference relative to the critical political situation in Germany.

The Kapp "Putsch".

For many months there had been recurrent rumors of an impending Putsch or coup d'etat in Berlin. But, ever since the revolution of November 1918 Germany, and more particularly Berlin, had been rife with rumors of imminent riot and revolution, most of which had proven false, and so American and Allied circles placed little credence in the latest rumor.

In the early days of March the political situation, always complicated in those days, became unusually tense. The underlying causes were the dissatisfaction of the Right with the "important" coalition Government and the consequent agitation for new elections. The demands of the Right were rejected by the Reichstag. The Extreme Left, which was not represented in the Cabinet, also gave voice to its own discontent with political conditions. The press of every political persuasion teemed with inciting articles of one kind or another, and such was the situation throughout the political parties that some observers declared the time to be prosperous for a putsch, particularly from the Right.

While the American authorities, through various agencies, were kept in touch with political developments in unoccupied Germany, yet the news that a putsch had really been made in Berlin came as a great surprise, as it did to many people in Berlin. With two Marine brigades, hard-bitten troops who had formerly served in the Baltic Provinces, the Extreme Right had made a desperate attempt to seize the reins of government.

It seems that the National Government had been advised that a radically monarchistic clique in Berlin had been agitating for an overthrow of the Government by unconstitutional methods. Orders for the arrest of several persons implicated in the plot, notably Dr. Kapp, were issued. During the night of March 12, General von Luttwitz, commanding the troops in the Berlin district, gave up his command. Dr. Kapp could not be found. From Doeberitz (near Berlin) where the Marine brigades were concentrated, came an ultimatum to the Government, demanding the formation of a new

Cabinet, the replacement of Defense Minister Noske by General von Luttwitz and amnesty for Kapp and others whose arrest had been ordered. If the demands had not been accepted by 7:00 A.M., the Marine brigades were to advance on Berlin.

After having been in session nearly all night, the Cabinet rejected the ultimatum. Before morning, however, it had become apparent that the other troops in an around Berlin and the Security Police, who had been directed to take up defensive positions, could not be depended upon. They were withdrawn and at about 5:30 A.M. the leaders of the Government left Berlin in automobiles. They had not resigned, and their last official act before their departure was the issue to the laboring classes throughout Germany of an appeal for a general strike as a protest against the Putsch.

By seven o'clock on the morning of March 13 the troops of the new government had entered Berlin. The First proclamation announced Dr. Kapp as National Chancellor and Minister-President, appointed General von Luttwitz supreme military commander and Minister of Defense, and declared that "a new government of order, freedom and action" would be elected.

Thereafter events moved rapidly. The old government went first to Dresden and then to Stuttgart. The general strike was called and obeyed, for the greater part, throughout unoccupied Germany. In Berlin, foreign press correspondents declared it "the most complete general strike in history". Gas, electric and street car services were suspended and even the water was cut off until the "Technical Emergency Aid" came to the rescue, while newspapers could not be printed. Bolshevism reared its head, and rioting and plundering occurred in Berlin and many of the larger cities of Germany. Street fighting occured in Berlin and other cities, and the radicals of the left attempted to take advantage of such a rare opportunity The Kapp Government quickly found that it could not maintain its position, as it had been denounced by all parties except those of the Right which gave it but weak support, and particularly because the general strike had "broken its back".

Meanwhile negotiations had been in progress, and the coalition parties had agreed to hold general elections within a short period. On March 17, Kapp issued a statement that as the old government had "fulfilled the most important political demands" of the new, he considered his mission concluded. He resigned forthwith and he and his chief supporters fled the country.

The constitutional government then returned to Berlin and resumed its functions. But the radicals had learned the tremendous power of the general strike, and the Government found itself in danger of being "hoist with its own petard". It was soon realized that it was far easier to bring about a general strike then to end it.

Riots and street fighting continued to occur at various points in Germany. The Marine brigade returned to Doeberitz on March 13, engaging while en route in street fighting in the very heart of Berlin. Not until March 23 was the general strike officially at an end in Berlin, while by no means had conditions returned to normal in all other portions of Germany. Conditions remained serious in Saxony, Thuringia and Mecklenburg, and particularly so in the great industrial section of Westphalia, generally known as the Ruhr District. The Communists secured temporary control in the former districts while in the Ruhr what was known as the Ruhr Revolt broke out. This was a clearly defined attempt to establish a government of the proletariat in Westphalia, and much hard fighting occurred before the National Government regained control of this important industrial region. (A description of the Ruhr Revolt is reserved for another chapter.)

Repercussions in the Rhineland.

The first news of the Kapp *Putsch* reached American Headquarters on the morning of March 13, through the medium of the Coblenz newspapers, including the Amaroc News, which carried the information in their regular morning edition. The American Intelligence officers, while taken by surprise were yet convinced that the long expected crisis was at hand. The first information was but meager and to the effect that the "Baltic troops" had entered Berlin and overthrown the Government, which had fled in automobiles.

Before informing Washington, endeavor was made to confirm the press reports Coblenz was alive with wild rumors, but the population remained calm. Not for several hours could telephone communication with the American Mission in Berlin be established and confirmation of the *Putsch* secured. This information, together with the names of the two leading spirits and a resumé of the more plausible rumors, was immediately cabled to the War Department. It soon developed that quick telegraphic communication between Berlin and Washington was impracticable, and thereafter the American Headquarters became the intermediary in forwarding dispatches to and from the American Mission in Berlin and the State and War Departments in Washington. An interesting feature of this critical period was that the American Headquarters in Coblenz was able to keep in constant telephonic communication with the American Mission in Berlin, communication having been established as many as five times in 24 hours.[1] Despatches came through the American Embassy at Paris to the American Headquarters at Coblenz, whence they were telephoned to Berlin, while despatches from the American Mission in Berlin were telephoned to Coblenz, they were enciphered and sent by telegraph through Paris to Washington.

While Coblenz remained outwardly calm, the popular mind was, of course, in a state of restrained excitement. As the details of the putsch and of the widespread disorder throughout unoccupied Germany came to be known, the Germans in the peaceful occupied territory realized that after all an enemy occupation had its advantages. Many residents of the American area expressed their thankfulness that the presence of American troops prevented the spread of disorder to that section of the Rhineland.

1 Once during a conversation, the communication was cut off by some operator located between the two cities, with the statement that telephonic communication with Berlin was prohibited. When told that an American officer in Coblenz was speaking, the operator said, "Ah! that is another matter", and communication was immediately reestablished, continuing uninterruptedly thereafter.

The "general strike", so widespread and so effective in unoccupied Germany, was comparatively abortive in the American zone and but little worse in the other occupied areas. Some Coblenz merchants boarded up their show windows and made other preparations for the disorder which they feared would occur upon the calling of the general strike, but these proved to be unnecessary precautions.

A general "protest" strike in the whole of the occupied territory, to last 24 hours, was called for March 15. At the same time it was stated that all railway trains and public utilities needed by the occupying forces would be continued in operation. The labor leaders instituted negotiations with the American and Allied authorities, relative to the securing of permission to call the strike. They were told that as long as the requirements for the maintenance and safety of the troops of occupation were provided for and no public disorder occurred the occupying authorities would not interfere. A large number of railway employees ceased work, but more than enough remained to insure the operation of important trains. In the American area the workers in a number of industries walked out, while in Coblenz the street car men and the waiters and musicians in most hotels and restaurants took part in the strike. Many of these returned to work before 7:00 P.M., and by the next morning (March 18) the protest strike had come to an end.[1]

1 Much amusement was caused at American Headquarters by the pleas of some street car men that the military authorities order them to resume work; a request which was, of course, refused. Striking waiters stationed themselves at the entrances to their places of work and petitioned passing American Officers for orders to go back to their labors. One young officer, without authority, gave the desired order which the striker and his comrades obeyed with grateful alacrity.

In the other occupied zones the strike was somewhat more general, but nowhere was there any interference with the needs or desires of the forces of occupation.

The "general strike" began and ended without the occurence of any real disorder in the occupied territory. There were large demonstrations in Cologne, Mainz, Aachen (Aix-la-Chapelle) and other cities and towns in the various zones occupied by the Allied forces, but they passed off without any untoward incident. The American area remained entirely quiet, and there were no demonstrations, probably because of a steady rain during most of the day. As has been pointed out in previous volumes, the inhabitants of the region which was designated as the American zone are by nature an orderly and law abiding people, and their placid minds are much less susceptible to incitement than are those of the population of industrial districts such as were comprised within the sections occupied by the other armies.

Naturally, all parties held frequent meetings to discuss the new developments and to decide upon their attitude toward the Kapp "government". While political discussion was heated and unending, the population declined to become excited and, secure in the presence of the occupying troops, adopted a "wait and see" attitude. At the same time the great majority freely expressed their views. With the exception of those who adhered to belief in a monarchical form of government. denunciation of the Putsch and a corresponding wish for the early return of the constitutional government were general. Some of those affiliated with the Extreme Right (National

People's Party) announced their pleasure in the overthrow of the former government, but in general all those of monarchist persuasion were content to hold their peace and await developments. Very few of any political belief thought that the Putsch would prove successful. Similar conditions prevailed throughout the Rhineland, and so far as public utterance went there was practically no one to mourn when Kapp and his aides departed from Berlin.

The officials of the Government of the Rhine Province, which is located in Coblenz, took no official action in regard to the Putsch. The Oberpraesident and his chief assistant were non-committal. Other officials were willing to express their attitude, but only in private conversation. Only a very few throughout the area expressed themselves as being in favor of the Kapp Cabinet, the great majority taking the stand that the move was extremely ill-advised and would but serve to bring further troubles upon an "almost intolerably harassed country". Fearful that the labor parties would demand a statement as to the position of the Government of the Rhine Province, the higher officials met in executive sessions to discuss the matter. Before a public statement became necessary, however, the collapse of the Putsch relieved them from an embarrassing situation.

The events in unoccupied Germany had little immediate effect upon the economic situation in the occupied territory, while the "protest" strike in the Rhinelands was not sufficiently widespread or long-continued to cause any marked disturbance of existing conditions. The supply of food and fuel continued to fall short of normal requirements and these shortages, combined with a lack of raw material, served to keep the industries in the occupied areas far from operating at full capacity. While the mark continued to increase in value, prices in general mounted to greater heights, and altogether the economic situation in the American area remained unsatisfactory.

During and after the brief existence of the Kapp regime, American Headquarters at Coblenz might well have been entitled, in war-time parlance, a "center of information". A constant and voluminous stream of information flowed into the "Regierungsgebaude", the government building which since December 1918 had housed the larger portion of the headquarters of the American Forces on the Rhine. It is of interest to note the various sources which contributed toward keeping the American Commanding General abreast of events. First there was the Intelligence Section of his own General Staff, with its trained operatives and confidential agents, which maintained constant liaison and continually exchanged information with the Intelligence services of the Belgian, British and French forces of occupation, and operated similarly with the American liaison officers at the other three headquarters on the Rhine and with American military attaches at the various embassies and Legations in Europe; in addition that Section maintained constant telephone. telegraph and courier communication with the American Mission in Berlin and the American Embassy in Paris, besides securing confidential information from official German sources. Then there was the Inter-Allied Rhineland High Commission, with its seat in Coblenz; the various departments of this body had their own Intelligence agencies and furnished American Headquarters with a mass of information -- not always accurate or without a tinge of propaganda, however, as was likewise true of the information received from all sources other than American. With all these and through frequent conferences with other military commanders, and considering the fact that the American representation on the Rhine and the United States Government had no selfish ends to pursue, General Allen was probably as well equipped as any man in Europe to form a correct and impartial opinion as to the events and issues that were potentially capable of bringing about another conflagration on the Continent.

Such, it would seem, became the opinion of our Government. He was called to Paris by the American Ambassador for a conference relative to the situation in Germany and to the attitude of our Government thereto, and afterwards, upón request, frequently submitted his matured views to the Ambassador in order to assist the latter in the performance of his high duties as a member of the Council of Ambassadors. It may well be surmised that General Allen had a part in determining the stand taken by the United States with regard to the Central European situation in the spring of 1920

It was not easy to form and maintain an impartial opinion. Every country concerned, including Germany. showed its keen desire to have the United States take its particular view of the situation, and to this end furnished the American Headquarters with a vast amount of information which was very often highly colored with propaganda according to the policy of that country. A conflict of views between the British and French developed almost immediately. The French made quick use of the story that the British had helped to bring about the Kapp *Putsch*, had promised early recognition of the new government, and had withdrawn their support only when its failure was seen to be inevitable.[1] In private conversation French in high position denounced "perfidious Albion" and declared that Britain's policy was to strengthen Germany at the expense of France. This was but one phase of the ultimate policy of England, which was to wreak harm upon the United States, her chief commercial rival; the farsighted English would pursue the traditional British policy towards their most powerful competitors, and history would again repeat itself. The British in turn claimed that France was so blinded by fear and hatred of Germany that what she saw in the Kapp *Putsch* was not an impending return of the Hohenzollerns, as she declared, but rather a valuable opportunity for the furtherance of her cherished scheme of the partition of Germany. With disorder in Germany as a pretext, the French would first seize Frankfurt on the Main and occupy the Ruhr District. This would be but a prelude to the division of the German *Reich* into small and weak German states. The isolation of Prussia could then be accomplished, the final result being great economic and political injury to France's powerful neighbor. According to the British view, this policy was inspired not only by fear and hatred but also by Napoleonic dreams of the hegemony of Europe. It was pursued with regard to the interests

1 This story was reported to the American Headquarters by Americans in official position in Berlin. These stated that certain British officials in Berlin were implicated, but that nothing definite as to the British Government had been ascertained.

of France only, it was reckless, and it would inevitably result in an ultimate condition of chaos in Europe. The Belgians, largely perhaps because of their military pact with France, generally followed behind the French. The Germans, more ardently desirous than ever of securing the good will of the Americans and their Government, declared that they had nothing to conceal and offered to furnish the Americans with all information desired. They were anxious to prove the falsity of French claims that a return of the monarchy, with all which that would involve, was imminent or even planned, and it was clear that they had high hopes that the wished-for slit in the Entente was at hand.

Realizing that Coblenz, as the seat of the American Headquarters, was like an island round about which swirled torrents of propaganda and intrigue, the American commander saw that it was a time for clear thinking, impartial reasoning and avoidance of controversial issues or of any act or speech that might be construed as an alignment with any interested party. These views he impressed upon his officers as a vital part of the general policy of the American Headquarters. The American course of action was to welcome every item of information, from whatever source, to give it most careful consideration to study and to analyze, and finally to form opinions that were based solely upon the information available, regardless of the fact that the side in error might have been a war-time comrade or that the merit of a particular issue might lay with a people with whom America had but recently been at war.

A stream of well digested information was directed upon the War Department and the American Embassy at Paris and frequent "estimates of the situation" were similarly forwarded. It was stated in Washington that through its representatives in Berlin and Coblenz, the United States Government had probably had as clear and accurate a picture of the situation in Germany in the spring of 1920 as it was possible to secure.

During the particularly critical days following March 13, the American Headquarters received a telegram from an individual (thought to be an American) in Leipzig, claiming that a number of Americans and British were stranded there, that heavy fighting was in progress in the city and their situation was rapidly growing desperate, and asking that a special train be sent to bring them to occupied territory. After consultation with the American Commissioner in Berlin and further communication (by telephone while the fighting was going on) with Leipzig, it was decided to send a train as requested. The German railway authorities complied with the American demands and furnished a special train without undue delay. Rations and medical supplies were put aboard the train, which carried an American officer in charge, a train crew composed of American soldiers available for use in case of emergency, a few American soldiers (unarmed), and a German officer sent along for liaison purposes. The train succeeded, after some delay but without particular incident, in getting to Leipzig, but found that the Americans and British had already left by an ordinary train that had been dispatched to Cologne earlier in the day. The train returned to Coblenz without incident, while some of the Americans for whom it had been sent came on to Coblenz from Cologne and were taken care of by the military authorities.[1]

At one time the American Commissioner in Berlin requested that a special train be made ready to be sent to Berlin to bring Americans to the American zone, but later found it unnecessary. Some Americans, mostly women, finally came by ordinary train to Coblenz where they were billeted and taken care of. There had been no real shortage of food among official Americans in Berlin, as the Coblenz military authorities had succeeded in sending them a carload of rations before their supply had become exhausted.

The Spring of 1920.

The gradual transfer of power from the military to the civil authorities of occupation continued during this period, and by the end of June the American Department of the High Commission had taken over many of the functions of the office of Civil Affairs.

The three months passed very quietly in the American zone. At the beginning of April the Ruhr Revolt was in full swing and there were disturbances in many other portions of Germany. Political and economic conditions in Germany were serious, and Frankfurt and Darmstadt were occupied by the French on April 6. But the reaction in the Rhineland was only slight, and particularly so in the American

[1]It seems that the person who had telegraphed from Leipzig and requested a special train, was an Englishman. Asked why he had applied to the Americans at Coblenz rather than to the British at Cologne, he replied that at first he had applied to the latter, but had met with the reply that as the British military authorities were powerless in the matter his application had been forwarded to the British Embassy at Berlin. Said he, "I will publish all over England that while the British Army declined to act, the Americans cut red tape and came to our rescue".

area. The events were discussed in political meetings and in the press, but temperately and without inciting oratory, while the people in general pursued the even tenor of their way. The politicians were active in preparations for the general elections, but the population took only a mild interest, and the elections were held without disturbances of any kind. In unoccupied Germany, and to a slight extent in the other occupied zones the strike fever and unrest continued; in the American area the frequent wage controversies resulted in but few strikes all of a minor character, and the population in general remained entirely calm. May 1, the traditional day in Europe for radical disturbances passed without untoward incident. Parades and meetings were held with the approval of the American authorities - throughout the area they were free from objectionable features. More than ever was it apparent that only in the event of extraordinary circumstances would serious unrest ever occur in the American zone.

There were a few disturbances in the other occupied areas, but none was serious. June was characterized by food riots throughout unoccupied Germany; demonstrations of this kind were made in the Belgian and French zones but they could hardly have been called riots. Because of the arrest of certain agitators by the French, a general strike was called at Ludwigshafen in the French area; nothing serious developed and work was soon resumed. The awarding of the Monschau railroad (between Eupen and Malmedy) to Belgium caused a few strikes in that region, but these too were quickly ended. A sudden strike of waiters and other hotel employees occured in Wiesbaden on April 3. Beyond the inconvenience caused hotel guests the strike was of no importance, the strikers resuming work within three days.

By June 30, the food situation in the American area had displayed a substantial improvement. (This was largely true of the other zones, and the people in the Rhineland in general were better off in this respect than those in most portions of unoccupied Germany. The ration regulations continued in force, but the allowance of potatoes and some other components had been increased. The campaign against profiteers and those who attempted to smuggle food out of the area went steadily on, the American authorities co-operating in the actions against food-smugglers. There was still a shortage of sugar and fats, but in general the outlook was better than it had been since the beginning of the occupation. Food was received from other countries, including the United States, and it was not so much a question of the amount of food supplies as of the prices, which continued to be almost prohibitive to the majority of the inhabitants of Coblenz and the larger towns. As compared with the previous year, the food situation was satisfactory, but it continued to be a problem for the larger part of the urban population. Considerable foreign aid was extended to the German people at this time, and in the American area an American Committee kept several kitchens in operation, feeding a large number of children.[1]

There was also an improvement in the fuel situation. It could not be called satisfactory, however, except by comparison, as the amount available was never sufficient to permit industries to operate at full capacity, and public utilities could not accumulate reserves in the required amounts. With the arrival of warm weather, the problem of the householder was simplified, and by the end of June the

1 Volume II, Chapter 8.

situation had become better than for months past, although the securing of stocks for the winter was still doubtful.

From 69 to the dollar at the beginning of April the mark advanced to 29 on May 26, declining to 38 by the end of June.

This rise in value resulted in the cancellation of many orders from foreign countries, causing a sharp decline in German export trade. This development, together with the numerous strikes, the higher wages secured by the workmen, and the resulting increase in the cost of living caused a business depression throughout Germany. But in the American zone the depression was comparatively slight. From time to time a few industries suspended operation because of a temporary shortage of coal or raw material. But in almost every case operation was resumed within a short period, and not at any time did the number of unemployed constitute a problem of any consequence.

The Rhineland separatist movement was not active during this period, although some of the newly formed Christian Peoples' Party (an offshoot of the Centrum) announced their adherence to the Dorten program. After the general elections party politics were overshadowed by the economic situation, and the period ended in a condition of political calm.

The rise in prices and in the value of the German mark had its effect upon the members of the occupying forces. At the same time that a "purchasers' strike" was in nation-wide vogue among the Germans, the Americans and other foreigners in the Rhineland were refraining from all but the most necessary purchases. This caused some reductions in the prices charged by a few of the Coblenz merchants, as it did in other portions of Germany, but the effect was not lasting.

In the endeavor to locate industrial equipment that had been removed to Germany during the War, several Belgian and French subcommittees of the Reparations Commission visited the American zone. With the specific approval of the American authorities and accompanied in every case by an American officer, they inspected various industrial plants, and in a few cases succeeded in their endeavor. The equipment discovered was either sent back to its original owners or they were reimbursed and the German plants permitted to retain it.

On May 30, Memorial Day was celebrated in many places. Detachments were sent to Paris and Romagne-sous-Montfaucon, and General Allen delivered the principal address during the ceremonies at the Romagne cemetery. Exercises were held at the cemeteries in Coblenz and Trier, and every American grave was decorated.

On June 2, his resignation having been accepted, Mr. Pierrepont B. Noyes gave up his duties as American Representative with the High Commission, taking with him his Deputy, Mr. Wallace H. Day. By agreement between the State and War Departments, General Allen, in addition to his military duties, took over the office vacated by Mr. Noyes. Colonel D. L. Stone, G.S., was selected by General Allen to perform the duties of Deputy American Representative.

Among the prominent visitors during the period were Hon. Ellis Loring Dresel, American Commissioner at Berlin; General Michel, commanding the Belgian Army of Occupation; General Degoutte, commanding the Allied Armies of Occupation; Hon. William Phillips, American Minister to Holland; General Rucquoy, the new commander of the Belgian Army of Occupation; Major General Peyton C. March, Chief of Staff of the United States Army; Major General John L. Chamberlain, Inspector General, U.S.A.; and Major General John F. O'Ryan of New York.

General March made a very thorough inspection of the A.F.G., covering about ten days. After the completion of his inspection, he expressed his high commendation of the excellent condition of the command.

The visit of General March caused fresh rumors of the withdrawal of the American troops, and again did the Germans voice their hope that the Americans would remain.

The Ruhr Revolt.

After the collapse of the Kapp *Putsch* the difficulties of the German Government were by no means at an end. The gravest problem, and the one of most interest to the authorities of occupation was the revolt in the Ruhr District, the edge of which was just across the Rhine from the Belgian and British zones. Almost wholly industrial, the Ruhr District had always been a trouble center, and the radicals there were quick to take advantage of the opportunity afforded them by the ill-advised attempt of Kapp and his associates.

In common with other portions of Germany, a general strike was called in the Ruhr District on March 14. Conditions became very disturbed and there were a number of small conflicts between the workmen and the police, as well as attacks by the former upon the Reichswehr troops and the Security Police. When, after the fall of the Kapp regime the general strike came to an end in most parts of Germany, it was continued in the Ruhr District. Apparently the more radical elements believed that the opportunity was at hand to establish a Soviet government or a dictatorship of the proletariat. The continuance of the strike seemed to be against the wishes of the majority of the workmen, but the others had their way The withdrawal of the Government troops was demanded; as compliance was not immediately forthcoming, Workmen's Councils were formed, many of the Security Police were disarmed, and in most of the cities and towns the municipal police were suspended from duty, their places being taken by Workmen's Guards which had been formed for the purpose. In some cases the constituted authorities were entirely suspended from office, but in most instances they continued to function under the supervision and control of the Workmen's Councils. A Central Workmen's Council was located in Essen, but it seemed never actually to exercise the power implied by its title; the headquarters of the movement was said to be in Hagen, and the names of the leaders were kept secret. In the meantime order was preserved throughout practically the whole district.

A demand by the commander of the troops stationed in the Ruhr District that the Workmen's Guards be disarmed, was immediately followed by combats between the workmen and the troops. The former proceeded to organize the so-called "Red Army" which was to drive the government troops from the district. This force was distinct from the Workmen's Guards which were to be used solely for keeping order in the towns. The Red Army idea proved very popular, a great many workmen joining it. A recruiting campaign was inaugurated, and many young men, not only from the Ruhr District but even from the occupied territory, were induced to join. Needless to say, many of the recruits were of a type that had no respect for law and order.

The Red Army then attacked the Reichswehr troops and the Security Police and after savage fighting drove them from the area, some taking refuge in the Belgian and British zones. This gave the workmen absolute control of practically the whole Ruhr District, and the Red troops took up positions from which to repel any attempts of the government forces to advance. It may be said that the Red Army was an army in name only and was loosely organized, in most cases the company being the largest unit. Their arms were secured by capture or from the homes of the workmen, and consisted largely of rifles and machine guns. only a few trench mortars and field guns being in their possession.

While the fighting was going on the Workmen's Councils held the reins of government, and forcibly suppressed any attempts at resistance. Except in the localities where fighting was in progress, order was maintained and conditions were fairly normal.

Negotiations with the Berlin Government resulted in an agreement in which many concessions were made to the workmen. Most of the Workmen's Councils wanted to comply with the terms of the agreement, but the leaders of the Red Army declined to acknowledge any authority higher than their own. The leaders then disagreed among themselves and the "Army" commenced to disintegrate into groups, some of which wanted to fight and others to return home.

A new Cabinet had been formed in Berlin, and on March 29 it sent an ultimatum to the Central Workmen's Council at Essen, requiring that all Red troops be disarmed by noon on March 30 and promising amnesty to all who complied. The leaders of the Red Army declined to comply, threatening to flood the mines and blow up the industrial plants. Negotiations were resumed and the time-limit was extended. The extreme radicals called another general strike; this was opposed by the great majority of workmen, but with armed force the radicals prevented the better elements from continuing to operate the mines and other industries.

Numbers of the Red troops had already commenced to return to their homes, and now dissatisfaction with the radical leaders became general. Workmen gathered in large crowds, denounced the leaders and demanded their overthrow - "We would rather have the Reichswehr in our towns than to have crooks and criminals who endanger the lives of honest men". The better elements of the Red troops returned to their homes, turned in their arms, and became normal citizens. Then the lawless element, in bands of from ten to several hundred men, drifted away from the "front", and a period of pillaging and looting set in. The new Workmen's Guards, which had been formed under the terms of the agreement with the Berlin Government, were unable to cope with the situation and appeals for help were sent to the Reichswehr, which in the meantime had commenced to advance.

The last phase of the Ruhr Revolt witnessed the advance of the Reichswehr and the restoration of order. Little serious fighting developed, the combats being mainly with small bands, with "mopping-up" operations as the final feature. Essen, the stronghold of the Reds, was occupied on April 6, and large numbers of Reds took refuge in British territory on the same day. By April

15 the Ruhr Revolt had been entirely suppressed, and by the end of the month, the disarmament of the workmen, while not complete, had reached a point that was generally satisfactory to the authorities.

Reaction in the Occupied Territories.

In so far as the German inhabitants of the occupied territories were concerned, the reaction to the Ruhr Revolt was but slight, and again they had reason to be grateful for the security afforded them by the presence of the troops of occupation.

The authorities of occupation took precautions to prevent the spread of disorder within the occupied territories, but it was soon seen that there was little danger of such a development. Some Spartacists, so-called, entered the village of Weisdorf in the British zone, displaced the local authorities and proclaimed a Soviet government. The leaders were promptly arrested by the British and the local authorities resumed their functions. The Reichswehr troops and the Reds, who at different times, came into the occupied territories (Belgian and British zones) were disarmed and interned; after several weeks, in each case, they were permitted to return to unoccuppied Germany.[1] While the revolt was at its height an officer of the Security Police who had participated in the fighting came to the American Headquarters for the declared purpose of buying tanks and armored motor cars from the A.F.G.; he was referred to a British syndicate that but recently had bought the large amount of surplus motor transportation left by the third

[1] Volume I, Chapter 2.

Army; there he succeeded in purchasing some motor transportation, but no tanks.

The authorities of occupation kept in close touch with the progress of the revolt, and again a conflict of views developed. The French took the stand that the revolt had been prevoked by reactionary German elements and that the concentration of troops for its suppression might involve danger to the security of the occupying forces. The advance of the Reichswehr troops, brought them into the neutral zone, and the French claimed that the number authorized by the Treaty of Versailles to be employed in the neutral zone had been exceeded. The German Government did not deny this, but stated that the excess was small in number and that it would be impossible with less troops to suppress the revolt.

Around this point the discussions among the occupying authorities revolved for some weeks The British and Americans held that the German Government had the right to use such force as was necessary to put down a revolt which imperilled its existence, and further that it was to the interest of all constituted European governments that without delay order be restored throughout Germany; the Ruhr District was the storm center and the German Government should be given as free a hand as possible in its efforts to bring about the return of normal conditions in that region, even if the authorized number of troops in the neutral zone should be exceeded. In this view the Belgian High Commissioner concurred. Each of the occupying forces made its own estimate of the number of troops in the neutral zone, but no two agreed.

The Americans, following their usual custom, sent operatives, including officers, into the Ruhr District to make independent investigations. From their reports and from a study of information from Belgian, British, French and German sources, the conclusion was reached that the German Government was employing no more troops than were necessary and that they exceeded the number authorized for the neutral zone by but a few thousands. The French estimate was far greater, while the Belgian and British estimates came between the French and the American.

As during the Kapp Putsch, the American Headquarters kept the War Department and the American Ambassador at Paris constantly informed of developments. There was frequent communication between the latter and General Allen who was several times called upon for his considered opinions. On April 5 the Chancellor (a position similar to that of Prime Minister) of the German Government forwarded to the American Commander the following declaration, with the request that it be transmitted to Washington:

> The German Government obliges itself to withdraw all troops in excess of the authorized strength as soon as operations in the Ruhr basin for the reestablishment of the Constitutional power have been concluded. The German Government has already delivered, on the 4th of this month, in Paris, the declaration that it expects to be able to do so within one week.
>
> Mueller.

This communication was immediately sent by telegraph to the War Department and to the American Ambassador at Paris.

Some time before this the German Government had requested the Allies to grant permission for additional troops to enter the neutral zone in order to put down the revolt in the Ruhr. While the other Powers were inclined to grant the request. France stated that she could permit such action only if she were allowed to occupy Frankfurt, Darmstadt and nearby towns as a guarantee that the additional troops would evacuate the neutral zone as soon as the emergency had passed. Germany rejected the French proposal, and on April 2 informed France that additional troops had entered the neutral zone without the consent of their Government; it was requested that the French give their consent thereto. The French Government declined this request and asked that the additional troops be withdrawn at once. On April 3, France dispatched a sharp note to the German Government, and at 4.45 A.M. on April 6 French troops occupied Frankfurt and Darmastadt, later taking possession of Homburg and Hanau.

The occupation was affected with opposition. The German troops had withdrawn, but a battalion of Security Police was captured in the Frankfurt barracks. There were two minor encounters between small Cavalry patrols; otherwise the movement proceeded according to schedule. During the afternoon of the next day French Morocean troops in Frankfurt, becoming nervous in the presence of a large crowd, fired on it, killing 6 and wounding 35 persons. Frankfurt was in a state of excitement for several days, but soon thereafter conditions became tranquil.

Except in the French zone there were no outward manifestations of excitement in the Rhineland. A general strike was called in Wiesbaden, but it failed to materialize. A state of siege in the Mainz bridgehead was proclaimed on April 6 and was lifted on May 2.

The occupation of Frankfurt raised a storm of protest in Germany where it was claimed that the move was preliminary to further advances into Germany. The French announced that they would withdraw as soon as the additional Reichswehr troops had evacuated the neutral zone, but the Germans claimed to believe that the French had no such intention.

On April 8 General Allen, accompanied by his Intelligence Officer, proceeded to Mainz and conferred with General Degoutte and M. Tirard, the French High Commissioner. The situation was thoroughly discussed and the French viewpoint was clearly explained to General Allen, who gave it careful consideration before reaching his conclusions.

Belgium soon showed her support of the French move by sending a battalion to participate in the occupation of Frankfurt. Neither Great Britain nor the United States supported the French and their disapproval, while tacit, was none the less apparent.

It had often been said that for a long time France had visualized the occupation of the Ruhr District. This view was supported by the arrival in the Rhineland of two French divisions, a number of additional infantry regiments, heavy artillery, aviation, etc. The French stated that this large force had come from France as a precaution against a possible hostile reaction on the part of Germany. Others claimed that there was no danger of such a reaction, and pointed out that the location of a large portion of these troops in the region west and northwest of Bonn was convenient for an advance on the Ruhr basin. It was generally believed at the time that if the disapproval of the British and

American Governments had not been apparent, the French would have moved into the Ruhr.

On May 17, investigation having shown that the German troops in the neutral zone had been reduced to within the authorized strength, the French troops evacuated Frankfurt and the other towns and returned within the Mainz bridgehead. The withdrawal was without incident, and shortly afterward the troops which had come up from France were sent back to their permanent stations.

An interesting incident of this eventful period was the defining of the status of the American troops of occupation. Late in March the House of Representatives by resolution requested to be informed, among other things, as to

> The extent of the authority exercised over American military forces now stationed in German territory by Field Marshal Ferdinand Foch, Commander-in-Chief of the Allied forces now in the occupied Rhine Provinces, and how far their activity may be directed without express orders from the President of the United States.

The President replied as follows:

> Replying specifically to the remaining questions in the resolution of the House of Representatives, I will state that Field Marshal Ferdinand Foch has no authority over the United States troops in German territory, nor can any one direct their activities without express orders from the President of the United States. It should be stated further that under his general police powers, under the terms of the Armistice, General Allen has full authority to utilize his troops for the police of the occupied district, the preservation of order, and to repel any attack which may be made upon him.

The resolution and the reply of the President was sent on April 5 to General Allen, "for his information and guidance". From the beginning he had followed the policy outlined by the President, but his stand thus became one that could not be questioned and it was clearly shown that he was independent of any authority other than his own Government.

July to September 1920

As compared with the preceding six months, this period was one of comparative calm in Germany. The Spa Conference in July, where Germany, under the threat of an occupation of the Ruhr District, agreed to deliver two million tons of coal per month to the Allies and was directed to expedite disarmament and the reduction of her military forces, was a bitter disappointment to the Germans. The award to Belgium of the two small plebiscite districts of Eupen and Malmedy, was another disappointment, somewhat softened by the overwhelming German victory in the plebiscite in East Prussia. The fighting in Poland and the seemingly imminent capture of Warsaw by the Bolshevist forces caused considerable tension in central and eastern Germany - which was relaxed after the French-inspired victory of the Poles. Fresh disorders occurred in Upper Silesia where there were frequent clashes between the Polish and German elements and occasional cases of friction between the Germans and the French troops in the plebiscite district. A number of communist demonstrations occurred in unoccupied Germany and rumors of a putsch, as well as of a secret Red Army in process of organization and training, were spread abroad. But, by comparison with the exciting spring months, the summer of 1920 was a period of calm and in the prevailing sense of internal politics was rather dull.

By the end of September all German troops had been withdrawn from the neutral zone, being replaced by the Security Police, and the Army had been reduced to 150,000 men.

The German inhabitants of the occupied territories were patriotically indignant over the outcome of the Spa Conference, they took a mild interest in the events in Poland and they shared the feelings of their compatriots relative to the happenings in the various plebiscite districts. As with their brothers in unoccupied Germany, foreign and domestic politics were pushed into the back ground by the increasingly difficult economic situation. There was a revival of the separatist agitation in the Rhineland, but it failed to make appreciable progress.[1] The arrest of Dr. Dorten in Wiesbaden[2] created a ripple of interest, but the summer was politically as dull in occupied Germany as it was in the unoccupied portions.

By the end of the summer the food situation throughout Germany had shown a substantial improvement and prices had decreased, although not in proportion to the amount of food in the country. The releasing of most articles of food from Government control had afforded opportunities for profiteering, and although there was plenty of food, except sugar and occasionally fats, the prices remained too high for the average consumer. During July there were more food demonstrations, several occuring

[1]The separatist movement had now taken the form of agitation for separation of the Rhineland from Prussia, but not from Germany.

[2]Volume I, Chapter 2.

in the French zone. Prices in the occupied territories, particularly in the American area, were said to be higher than in unoccupied Germany, and statistics showed that food was cheaper in Berlin than in Coblenz. American foodstuffs were available for purchase, but the prices were prohibitive to Germans. Large harvests of potatoes and grain were largely responsible for the improvement in the food situation, but at the end of September prices were again on the upward trend, and it was found necessary in the American area to include more districts in the distribution of food to German children.

The fuel situation in Germany became serious during the summer and was not less so in the occupied territories. The shortage of coal was due to the coal deliveries made under the Spa agreement, to the usual summer decrease of output, and to the strikes and other troubles in the Ruhr District, the Saar basin and Upper Silesia. In common with other portions of Germany, many industries in the American zone were forced to suspend operation or to discharge portions of their employees. Sufficient fuel for public utilities was secured, but neither industries nor householders could be supplied for their present needs, to say nothing of the coming winter when, it was feared, an acute shortage would develop.

Towards the end of the summer the business depression became somewhat lighter. The mark had fallen from 38 to the dollar on July 1 to 62 on September 30; this resulted in more foreign orders and the ability to secure more raw material, but up to the end of September the beneficial effect had not been great. The entire closing down of many industries and the part time operation of others resulted in a large increase in the number of unemployed in the American area as well as in other portions of Germany. The number in the American zone was added to by an influx of unemployed from other sections. These came principally to Coblenz, where the police immediately became active and deported most of them. The unemployment situation was never serious, and in general industrial and business conditions in the American area were better than in any other portion of Germany, and, despite high prices, its laboring class remained free from unrest of any consequence.

Labor disputes and strikes continued, but none of a serious nature occurred in the American zone. In July, as a protest against the high food prices, a general strike was called at Kreuznach[1] in the French area. The merchants promptly made a 50% reduction, which later by general consent, was changed to 20%. The most serious strike occurred in the Saar District where a general strike was called on August 7. The strike was political rather than economic in nature, and was called as a protest against the control exercised by the Saar Commission, the body that governed the district under the authority of the League of Nations. The French declared a state of siege, rushed troops to the scene and prevented any but minor disorders; their strong steps resulted in most of the strikers returning to work within two days. By August 14 the strike was entirely over, and the state of siege was lifted on that day. A number of mining officials were dismissed for participating in the strike.

[1]For a long period during the War Kreuznach was the seat of the German G.H.Q.

As a result of radical agitation in several localities in the French area, the Kreis representatives of the High Commission were empowered for a period of two months to prohibit meetings, assemblies and night traffic whenever such action seemed advisable. The measure was applied at Ludwigshafen and evoked sharp protest from the German press which claimed it to be unnecessary.

July 4 was celebrated in all garrisons in the American zone. The largest celebration was in Coblenz where the salute to the Union was fired at noon. In the morning there were games and sports of various kinds, followed by baseball games and championship boxing matches in the afternoon. The celebration was ended with a fireworks display from Fort Ehrenbreitstein, which almost equalled that of the previous year when the great stores of German rockets and flares, which had been surrendered as enemy war material, were used. During July both the French and Belgian national holidays were celebrated by the Americans with salutes and fitting exercises.

Two events of interest during the summer were the conference of American military attaches, held at Coblenz in July, and the September maneuvers in which all combat troops of the A.F.G. participated.

The most distinguished visitors during the period were President Millerand of France and Marshal Foch who arrived in September. The Commanders of the Belgian, British and French forces were guests of General Allen on the same day.

Among other visitors during the summer were Colonel M. Thompson, Senator Walter E. Edge of New Jersey, Senor Delgado,

(Spanish Minister at Berlin), Senator Henry L. Myers of Montana, Representative Harold Knutsen of Minnesota, and Bishop Brent (formerly Chaplain-General of the A.E.F., and well known to all Army officers who had served in the Philippines).

During this period General Allen exchanged visits with the Commanders of other forces on the Rhine and made two official journeys to Paris.

By the end of September the American Department of the Rhineland Commission had taken over all the duties contemplated under the ordinances. With this the period of military supervision and control of the German civil government in the American area came practically but not entirely to an end. The American military authorities continued to exercise a certain control over the German authorities having to do with billeting, requisitions, coal, public health, etc.[1]

The Last Quarter of 1920

These three months constituted what might have been called a period of party conventions, but domestic politics were overshadowed by economic conditions and gloomy forebodings of Germany's future. The new Prussian constitution allowing more power to provincial governments was adopted in October, giving an impetus to the various separatist movements in that state. The "away from Berlin" movement for the complete autonomy of Bavaria was again to the fore; it subsided when in December the tension between Germany and the Allies, relative to disarmament and the disbandment of self-defense organizations, caused

[1]See Volume II, Chapters, 3,5,6,9,10,13,14,21,22,24 and 25.

Bavaria to come to terms with Berlin, and the Bavarian Government affirmed its loyalty to the German Reich. There was comparatively little disorder in Upper Silesia, but the situation there was always delicate and the interest of the German people in the plebiscite (the date for which had not yet been set) and its outcome grew keener as time went on; autonomy was promised that province of Prussia, should it be allotted to Germany of which grave doubt was felt. There was much radical agitation and a recurrence of the rumors of impending attempts by Communists and monarchists to overthrow the government. The disarmament campaign was in full swing and great numbers of weapons of all kinds were turned in by "self-defense" organizations and individuals. Except in the matter of prices, which continued to rise, the food situation remained practically unchanged, but strikes were frequent, while the coal shortage and slump in the export trade caused hundreds of industries to suspend operation, resulting in a large increase in unemployment. The year closed with the German people in a thoroughly pessimistic mood, while the press teemed with assertions that France was bent upon the destruction of Germany. Only a few expressions of hope were heard, and those who made them said that Germany's only chance to emerge from the "fearful slough of despond" lay in the extending of a helping hand by the United States. The period was unquestionably a difficult one for Germany and the near future bade fair to be a time of trial, but it was none the less a fact that at the end of 1920, Germany was in better shape than she had been since November 1918. Then too, peace had come and Germany was in diplomatic relation with practically every country except the United States. By December 31, the German Army had been reduced to 100,000 men, which was the strength authorized for the permanent force.

The conditions of life were difficult for the people of the American zone also. The acute coal shortage not only caused many industries to close but involved much discomfort to householders and even suffering to the poorer classes. There had been a substantial decrease in unemployment in October but at the end of December it had increased by about 100%. Prices of food had undergone a further increase, and the problem was a serious one for the average German; meat was a luxury and the supply far exceeded the demand, - so much so in fact that while the cost of living was higher in Coblenz than elsewhere in Germany, meat became cheaper because so many abstained from its use except on rare occasions.

As elsewhere in Germany, the vicious circle of increasing cost of living and demands for higher wages continued in the American area. Hoarding by the farmers and profiteering by the "Schiebers" had much to do with the high food prices, and occasionally the townspeople took matters in their own hands. Sometimes they raided farms and forcibly caused the owners to dispose of their hoards of potatoes and grain, and sometimes crowds made peaceful demonstrations in the country districts and by direct negotiations induced the farmers to release their stocks at an agreed price. At Andernach the American authorities seized nine carloads of potatoes at the request of labor unions and caused them to be sold at a fair price. The German Government continued its campaign against food profiteers and hoarders, but by the end of the year the campaign had not succeeded in lowering food prices in general.

The very low water in the Rhine during most of this period had a share in the causing of the coal shortage. The shortage became so acute that at one time a number of blacksmiths in the American area had to close their shops. Restrictions as to the use of light and power were again imposed, but public utilities were never without reserves.

From 61 to the dollar on October 1 the mark had fallen to 89 on November 9. Then there occurred a sharp rise and on November 18 the dollar was worth only 60 marks. This violent fluctuation caused the cancellation of a great many foreign orders and so had a harmful reaction upon trade and industry in general. The sudden rise in value was entirely unexpected and was never satisfactorily explained, although it was generally believed that speculation was largely responsible. A week later the mark had fallen to 70 and it remained between 70 and 75 until the end of the year.

As in other portions of Germany, November 9, the second anniversary of the revolution of 1918, passed quietly in the American area. There were no demonstrations and the few meetings held were entirely orderly.

Except the politicians and the official class, the people in the Rhineland seemed to take but a mild interest in the many problems with which Germany was beset. The politicians grew busier with preparations for the Prussian elections, but the population remained almost apathetic in that regard. The press voiced much interest in the election of President Harding and in his pre-election assurance that he would withdraw the A F.G. soon after his inauguration, and expressed pleasure over the choice of the American people, together with hopes that the new Administration would give aid to Germany. The visit of Senator Medill McCormick in late December excited much interested comment; he was looked upon as the personal representative of the American President, and it was hoped that the former would take back to his chief "the true picture" of Germany.

After having been more or less dormant for some time, the Rhenish separatist movement came actively to the fore and there was a lively renewal of the agitation. By this time the separatists had formed themselves into three distinct groups. One planned to constitute the Rhineland and the Palatinate into a Rhenish federal state, autonomous and free from Prussia, but an integral part of the German Reich. Some of the Rhenish section of the Catholic or Centrum Party supported this idea. The Dorten movement was originally started with a view to the construction of the Rhineland, Palatinate, and portions of adjoining provinces into an independent republic which was to be a buffer state between France and Germany; the Dorten group now claimed to have discarded that idea and to be working for an autonomous state within the Reich. They were supported by the Christian People's Party. A third group, composed of the Rhenish People's Party, and the Republican People's Party, of the Palatinate, advocated an independent republic, entirely separated from the Reich and prising the Rhine Province, Rhenish Hesse, Birkenfeld, and the Palatinate The difficult economic situation and the amount of reparations, not fixed as yet but certain to be a heavy burden, were having their effects on the minds of the Rhinelanders -- separation from Germany, or even from Prussia only, might mean relief

The new Prussian constitution and the promise of autonomy to Upper Silesia inspired the separatists to fresh endeavor. There was no question but that the great majority of the German inhabitants of the occupied territories wished to remain within the German Reich; it was a fact, however, that a great many desired the Rhineland to become an autonomous federal state. In other words, separation from Prussia but not from Germany. At the same time the greater part of the population were not actively interested, but politicians and various newspapers, some established for the purpose, began a lively agitation of the subject. All parties took up the matter and there was a vast amount of heated discussion pro and con. The local factions of the important political parties were against complete separation, but there was so much agitation that for some time most of them evaded the real issue and failed to take a definite stand upon the autonomous state question. While the population declined to become excited, the Central Government become exercised over the matter and in November Chancellor Fehrenbach and Foreign Minister Stevens visited the occupied areas and made several speeches, assuring the Rhenish population of the Governments interest in their welfare and commiserating with them upon the burdens caused by the occupation. Nothing definite was promised, but hints that their wishes would eventually be granted were made Other Government officials visited the Rhineland and it was clear that the lively discussions in party meetings and in the press had caused Berlin to become worried.

The Rhenish Centrists were not a unit in their views relative to Separatism: some wanted a federal state while others would be satisfied with provincial autonomy as a part of Prussia. The Rhenish Provincial Diet met at Duesseldorf in December, but it failed to take a definite stand. The net results of the revival of separatism were the bringing of Rhineland matters forcibly to the attention of the Central Government and the partial clarification of the general subject. It was clearly shown that the movement for separation from Germany was supported by only a comparatively few people of little or no political prominence and, as well, that the movement for a federal state, separate from Prussia, was steadily growing stronger, despite considerable opposition.

The speeches made by the German Chancellor and Foreign Minister at Cologne and Aix-la-Chapelle had contained statements with regard to certain actions of the Allies and to conditions in the Rhineland that met with the disapproval of the Allied Governments. In early December the French and British Ambassadors and the Belgian Minister at Berlin delivered a joint note to the German Foreign Minister in which the three governments protested strongly against the speeches and pointed out that similar statements must not be made again.

During this period there developed among the German officials in the Rhineland an apparent policy of obstruction and opposition to the authorities of occupation, and it was found necessary by the High Commission to dismiss several of them from office.[1]

The labor unions in the American area engaged in frequent wage disputes with their employers and some strikes were called; none was of a serious nature and but two or three continued for any length of time. A number of strikes were averted through the mediation of the occupying authorities. The railroad and postal employees throughout Germany demanded increases in wages and threatened to strike if their demands were not complied with. At first policy

[1] Volume I, Chapter 2.

of "passive resistance" was adopted by the employees, but later the threats of a strike. Towards the last of December a strike ballot was taken - the railroad employees requested and received the permission of the occupying authorities to conduct the referendum in the Rhinelands. An overwhelming majority voted in favor of a strike and at the end of the year a general railroad strike seemed imminent.

With the approach of Christmas, the Amaroc News, the daily newspaper conducted by personnel of the A.F.G., instituted a campaign for contributions towards giving presents to the poor children of the area. Over 360,000 marks were contributed by the Americans, and in all garrisoned towns Christmas tree celebrations were held and presents were distributed to all children who attended. This action of the American soldiers evoked many expressions of appreciation from the German press and people.

Among the visitors at different times during the period were several officers of the Dutch General Staff, two officers of the Spanish General Staff, and the Chief of Staff and several other officers of the Swiss Army. They were made acquainted with the various activities of the A.F.G. and were afforded opportunity to inspect the troops and witness their methods of training.

Other visitors included Senator Medill McCormick of Illinois and Representatives E.S. Chandler of Mississippi, A.T. Fuller of Massachusetts and A.P. Patten of the same state. In December General Degoutte came and conferred upon General Allen the decoration and title of Grand Officer of the Legion of Honour.

Later in December General Allen made an official visit to Brussels where he conferred with the American Ambassador, the King of Belgium and high officers of the Belgian Government and Army. Shortly afterwards he went to Paris for conferences with President Millerand and the American Ambassador to France.

January to March 1921

The beginning of the year was marked by the acute political situation that developed in Germany in consequence of the ompending reparations demands, the pressure of the Allies with regard to disarmament and the troublous question of Upper Silesia, together with the Prussian elections Communist activity, the usual rumors of radical and reactionary coups, and the difficult economic conditions. The action of the Paris Conference in fixing the reparations caused a wave of indignation throughout the country and a united front against accepting the Paris decisions Popular feeling ran high and one result was the suspension of almost all internal political dissensions pending the outcome. The Allies' rejection of the German counter-proposals and the London ultimatum, followed by the occupation of the important Rhine ports of the Ruhr District and the Allied preparations for the application of the economic sanctions, all caused further indignation and an increased spirit of national unity. It was an eventful period, and while reparations matters were the overshadowing feature, the others served to keep public feeling in a state of tension. There was a lively exchange of notes on disarmament, and compliance with the Allied ultimatum caused some excitement in Bavaria, although that state took occasion to repeat her assurance of "no break with Berlin". Frèsh disorders occured in Upper Silesia, however, March 20, the date of the plebiscite passed quietly. The rejoicing over the German Victory (61% of the vote was coupled with grave fears that the Supreme Council would award a portion of the region to Poland. The Prussian elections proved a heavy blow to separatist movements and resulted in gains by both the radical and the reactionary parties at the expense of those of the middle. There was a marked increase in Communist activities and demonstrations were made throughout the country. Such disorders as occurred were of a minor character, except in Saxony; the Government used strong measures, and by the end of March it was clear that the extreme radicals had lost ground through their ill-timed and ill-advised actions. For a time the air was full of rumors of imminent revolution - the "Reds" were going to proclaim a Soviet republic the monarchists were about to seize the reins of Government and establish a constitutional monarchy, and both Right and Left had combined to throw the country into "National Bolshevism" - but nothing happened, and the end of the period found the people as a whole rather callous to rumors of revolution and more than usually determined that there should be no change in their form of government, in the near future at least. The national feeling was stronger than at any time since the revolution of 1918, as shown by widespread celebrations of the 50th anniversary of the founding of the German Federation.

The country passed through serious economic difficulties at this time. During the earlier part of the period the food situation had grown worse, largely because of the steady increase in prices, and industrial and trade conditions also had become more unfavorable. Financiers took a gloomy view of the situation, while the continuing coal shortage and the innumerable strikes had served further to complicate the already difficult economic problem. But in February conditions began to improve, and by the end of March the economic situation in general had displayed an appreciable improvement. In the beginning of January the German mark was quoted at 62 to the dollar. Its greatest value was reached on January 27, when the Dollar bought only 55 marks. The fluctuations during the three months were not great, and at the end of March the mark stood at 63 to the Dollar.

The Rhenish separatist movement received a decided setback during this period. The Rhinelanders in general had displayed indifference during the Prussian elections campaign, although the leaders of the various parties were very active. Aside from their ever-present economic difficulties, their interest was centered in the reparations and disarmament demands and the resulting developments. The people as well as the political parties were united in opposing an acceptance of the Allied demands, and it was said on all sides that domestic dissension must cease in order that the Allies be faced with the opposition of a united Germany. Even the separatist parties, with the exception of the Dorten group which split over the issue, announced that on account of the Paris demands they would cease their agitation and that "now all differences must disappear". Closely following this came the Prussian elections in which all separatist groups, including the Christian People's Party were decisively defeated. Not long afterwards the elections for the Rhenish Provincial Diet were held, and again the separatist parties suffered overwhelming defeat. In March the Diet met and by resolution showed itself to be almost solidly against separatism in any form. Under these heavy blows the separatist movement was forced into the background, and at the end of March was practically dormant, although the leaders hopes that the sanctions would cause interest in it to be revived.

While the inhabitants of the American area were deeply intereste in the reparations and disarmament matters and in the Upper Silesian question, they preserved their habitual calm and the period passed without any manifestations of excitement. They too were disappointed in the inaugural speech of President Harding - they had hoped for "help" from the United States and his failure to mention Germany in his speech was a distinct shock to their expectations.

The most interesting German visitor of the period was Noske, the Oberpraesident of Hanover. Until the spring of 1920 he had been Minister of National Defense and as such had put down lawlessness with an exceedingly strong hand, gaining for himself a wide fame and earning the hatred of the radicals. He made several speeches during his stay of several days and was remarkable for being one of the very few prominent Germans who at the time (January) voiced any optimism as to the future of Germany.

There was a noticeable increase of activity among the Communist elements in the Rhineland, more particularly in the Belgian zone where repressive measures by the occupying authorities became necessary. The French experienced some trouble and sentenced several agitators to heavy fines and long terms of imprisonment. There was less activity in the British and American areas but it was sufficiently noticeable to draw the close attention of the authorities of occupation. Five agitators were arrested in the American zone for having held a meeting without having given the required notice. They were sentenced to imprisonment, a step that evoked approving comment from the German press, but were released within a short time afterward. The movement was never serious in the American area and by the end of March had almost completely collapsed. That the movement failed so quickly in the occupied territories was affirmed by the Rhinelanders to have been almost wholly due to the presence of the troops of occupation. But an important contributing factor was the strong opposition of the great majority of the workmen who would have nothing to do with the extreme radicals.

During January there developed a press propaganda against the American troops because of their "luxurious" ways and their "responsibility" for the high cost of living in the American area. This occurred mainly in newspapers in unoccupied Germany, but in a few instances the local papers contained veiled attacks upon the American authorities. No attention was paid to the propaganda and it soon ceased.

Labor disputes continued and there were a number of strikes during the period. There was no disorder in the American area and, except for minor Communist demonstrations, none in the other zones. As in other portions of Germany, the Rhineland passed through a period of difficult economic conditions. The coal shortage was the most serious feature. At one time in the American area farmers offered to exchange their surplus potatoes for coal, but they could find no one able and willing to make such a trade. By the end of the period the fuel situation had improved to such an extent that most industries in the American zone were operating at greater capacity than for months past, although fears of the effect of the economic sanctions had caused the cancellation of many orders from unoccupied Germany.

In February, after reaching very high figures, food prices began to decline. The decline continued as more supplies became available, and by the end of March the food situation in the American area was by comparison nothing short of satisfactory. There was still a shortage of fresh milk, however.

A considerable decrease in unemployment occurred in the American zone during the three months. The German authorities were active in providing emergency work and the number of unemployed never grew to an important figure. Much construction work was in progress, mainly on residences and apartment houses to be used by those belonging to the A.F.G. or the High Commission, and there was little real difficulty in finding work: it was probable that a number of those registered as being without work were in reality of the professional unemployed type.

The general strike of railroad and postal employees which had seemed imminent at the end of the year, was averted in January by the act of the Government in granting an increase of wages and making other concessions.

The low water in the Rhine interfered seriously with navigation and was at least partially responsible for the decrease in coal deliveries to the Allies. At the end of March the Rhine was lower than it had been since the first of the year.

The 50th anniversary of the founding of the German Federation was not celebrated in the Rhineland as the High Commission had prohibited all meetings and demonstrations in that connection. Otherwise it would have been widely celebrated, for after all the majority of the Rhinelanders seemed to remain monarchists at heart and so to be lovers of the old order of things.

In January Field Marshal Sir Henry Wilson, Chief of the British Imperial General Staff, visited Coblenz for three days during which a review was held in his honor. A week later M. Albert Deveze, the Belgian Minister of National Defense arrived. He too was given a review during which he decorated General Allen and several of the latter's staff.

A delegation of the Franco-American society visited Coblenz in February. In March Generals Degoutte and Weygand[1] of the French Army came to Coblenz and conferred with the American Commanding General, and in the latter part of the month Colonel R. R. McCormick of Chicago, one of the owners of the Chicago "Tribune", was a visitor. During March a review of the combat troops of the A.F.G. was held in honor of the British High Commissioner, Mr. Malcolm Arnold Robertson.

The Sanctions.[2]

From the ratification of the Treaty of Versailles to the early part of 1921 the Allies had been pressing the German Government to comply with the terms of the Treaty, particularly with reference to disarmament. The Boulogne Conference in June 1920 had resulted in specific directions to Germany relative to the reduction of her military forces and the destruction of war material. The Spa Conference in July of the same year had forced Germany, over her protest, to agree to the demands of the Allies regarding disarmament and the delivery of coal, while Germany had been permitted to try German war criminals before her own courts and the reparations question had been postponed until a further conference to be held at Geneva in Switzerland.[3] While realizing that there were many difficulties in the way of Germany's compliance

[1]Chief of Staff to Marshal Foch. The dramatic defeat of the Bolshevist army by Polish troops in the summer of 1920 was generally attributed to his intervention.

[2] Volume I, Chapter 2.

[3] This conference failed to materialize.

with their demands, the Allies took the stand that she had been unduly dilatory and that further measures of coercion were necessary. France had dispatched several strong notes to Germany, calling her to account for noncompliance with the disarmament demands, including the disbandment of the so-called self-defense organizations, but the results achieved were far from satisfactory. Germany claimed to be doing all that was possible to fulfill the terms of the Treaty and of the agreements made since its ratification, and in the meantime pressed for a fixing of the amount she would have to pay as reparations.

As a result of the Paris Conference, held in January 1921, the Allies issued further directions to Germany relative to disarmament and informed her of the decision regarding the amount of reparations. Germany was to pay 226 billions of gold marks in 42 years and was to turn over to the Reparations Commission 12% of the value of her exports. The German Government was inclined to accept the disarmament terms, but at once announced that the reparations demands were not acceptable. An invitation to attend a conference at London was accepted with the proviso that Germany be allowed to submit counter-proposals relative to the Paris terms, which had included certain military and economic sanctions to be applied in case they were not accepted.

The London Conference met on March 1, 1921. The Germans submitted their counter-proposals which were rejected without discussion. The Germans were given four days within which to accept the Paris decisions or to submit acceptable counter-proposals; if neither course was adopted the Allies would proceed to apply the sanctions. Backed by all political parties of importance, the German Government stood firm in its resolve to reject the Paris terms. It submitted new proposals which were promptly rejected by the Allies, and on March 7 the Conference was abruptly terminated. In the meantime the Allied forces had been moving into position, and on March 8 French, British and Belgian troops advanced into the neutral zone and occupied the cities of Duesseldorf, Duisburg and Ruhrort, the Rhine ports of the great Ruhr--Westphalia coal and industrial district.

The French had made a partial mobilization in preparation for this movement and a large number of troops came from France for the occupation of the three cities. The Belgian contingent was much smaller than the French, while the British force was composed of a detachment of cavalry, a small amount of aviation and several tanks.[1] The advancing troops met with no opposition and the population preserved a correct and peaceful attitude. A State of seige was immediately proclaimed, but only minor restrictions were imposed.

A French general had been placed in command of all troops in the newly occupied territory, this caused the Belgian Government "keen emotion" as it was, the Belgians said, in violation of the agreement made at London. The matter was soon settled by allotting Duisburg and Ruhrort to the Belgians, leaving Duesseldorf to the French, and constituting the new territory a Belgian bridgehead under command of General Rucquoy, the Commanding

[1] The British contingent was withdrawn shortly afterwards.

General of the Belgian Army of Occupation.[2]

The German press and people failed to manifest much excitement over the application of the military sanctions, while the German Government withdrew its second proposals made at London, sent a note of protest to the League of Nations, and adopted a policy of "passive resistance", refusing to make the reparations payments that came due in March. The stand taken by the German delegation at London was enthusiastically approved in Germany, and interest turned to the forthcoming application of the economic sanctions.

An interesting sidelight on the London Conference was afforded by the reported statements, in private conversation, of Foreign Minister Simons who had been the leading German representative there. He said that the German delegation were optimistic and felt that the Allies had made a blunder. Germany had offered to pay for five years practically what the Paris Conference demanded; after that time further payments could be discussed and settled. He believed that had he, Briand and Lloyd George been able to decide the matter between themselves an agreement would have been reached. "But with Foch there, in that atmosphere and in the present state of French public opinion, a reasonable understanding was impossible. Lloyd George tried his best to compromise, but Briand was too strong, and the former was

[2]No bridgehead had been allotted to Belgium in the original distribution of the occupied territories, a "discrimination" that was felt keenly by the Belgians. In March 1921 a Belgian said: "Failure to give Belgium a Rhine bridgehead was an oversight at Paris in 1919. The new bridgehead will be held as long as that of Cologne". The latter statement was interesting in view of the supposed temporary character of the sanctions.

much disappointed over the rupture of negotiations".

The German Ambassadors at London, Paris and Brussels were ordered to Berlin, and it was rumored that they had been recalled. It was said that they had been called there simply for a conference with the Government, although the time of return to their posts had not been definitely fixed. Some time later they returned to their Embassies and the incident was closed.

In the American Area.

As in the other occupied territories the population of the American area remained entirely calm. For several days French troops of all arms passed through the area, some marching through Coblenz, but the Germans preserved an attitude of practical indifference. Their main interest was directed towards the form the economic sanctions would take and their effect upon industry and trade. The fact that the United States had not participated in the military operations evoked expressions of mild appreciation. The Germans had learned, however, that while the Americans might not participate in measures against Germany they would not become involved in a disagreement through which Germany could evade her obligations.

In his capacity as American Representative with the High Commission, General Allen was in frequent communication with the State Department and the American Ambassador to France, mainly with references to the form of the economic sanctions and the attitude of the American Department towards them. The interest of the Washington authorities was shown by the following dispatch sent on March 8 by the Secretary of State:

> Keep Department closely informed in connection with Allied advance in the German territory, of any military or civilian action or any ruling of Rhineland High Commission which might affect the United States and of any other matters of interest in your district.

Information of the military movement and of the economic sanctions decided upon by the Allies had already been cabled to Washington, and thereafter the State Department was kept constantly informed of developments in the organization of the customs system. In like manner the War Department was advised of all matters that should properly be made known to it, particularly those relating to the military situation and those affecting the American troops in any way.

As one feature of the economic sanctions was the establishment of a line of customs posts between the occupied and unoccupied portions of Germany, the question of the American area became of vital importance. Neither the Allies nor the High Commission could exert any authority in the American occupied territory without the permission of General Allen. It was foreseen that the United States would not participate in the economic sanctions. Unless there was made some arrangement whereby the customs barrier could be placed on the eastern border of the American zone, the economic sanctions could not be made effective. Were the American bridgehead to constitute a gap in the proposed customs line, not only would it render abortive the Allied plans but also it would present to the world an apparent evidence of disagreement that might well have far-reaching consequences. Acting upon the principles that in January of the previous year had caused him to make the ordinances of the High Commission effective in the American zone, General Allen on March 9 sent the following cablegram to the Secretary of State:

> If sanctions of London Conference be considered solely as penalties and temporary it would not be wholly inconsistent for me as commander of force of occupation under Armistice conditions to permit their execution in American Zone by High Commission. This refers only to Rhine bridgehead customs, as none of the customs posts of Rhine Province contiguous to Holland, Belguim and France are in American Zone. As delicate as our position is, it would be unfortunate for general welfare to make changes in our status now. The Commission fully appreciates our difficulties and is most considerate of my views. As in past I will make Commission's ordinances my orders as far as they are compatible with Armistice conditions.

The High Commission, which had been charged with the organization and conduct of the customs system in the Rhineland, forwarded its proposed scheme to the Allied Conference for approval. The delay in receiving this approval resulted in failure to commence the operation of the customs system until early in April, although before the end of March a preliminary organization was in operation in the Mainz bridgehead. In the meantime General Allen had informed the High Commission that customs posts could be placed along the perimeter of the Coblenz bridgehead; while the Americans would not participate in the conduct of the customs system, as Commanding General of the A.F.G. he would furnish guards at the customs posts, but solely for the purpose of preserving order.

On March 23 the Secretary of State indicated his approval of General Allen's policy in the following cablegram:

> In response to your various telegrams on the subject of Allied custom control in the American area, you are informed that although this government will take no part in the enforcement of penalties decided on by the Allies it does not

wish to put unnecessary obstacles in the way of such enforcement. This government therefore perceives no objection to the placing of Allied custom collectors at the request of the Rhineland High Commission in the bridgehead in the American Area, if you consider that this will in no way endanger the safety of the American forces of occupation nor interfere with the authority vested in you under Armistice conditions.
Secretary of War has been informed of this telegram.

As soon as he received information of the form of the economic sanctions, General Allen took steps to ascertain their probable effect upon industry and trade, including trade with the United States.

It was concluded that the erection of a customs barrier would probably have a harmful effect upon the industry and trade of the Rhineland and of unoccupied Germany as well, but not to the extent claimed by the Germans unless it should be long continued. The delay in setting up the customs system was injurious as business was "suspended in the air, waiting to learn the conditions under which it must work". As far as could be foreseen the proposed customs regime would not injure American trade with the Rhineland, aside from probable delays in deliveries and the possible effect upon prices of Rhenish manufactured articles and upon the general business and industrial situation in the occupied territories. The detailed report was forwarded to the State Department on March 24.

A description of the customs regime will be found in the chapter (No. 2) on the Interallied Rhineland High Commission and its appendices.

While it was generally understood that the sanctions were to be temporary measures, differences of opinion in this regard were apparent in the High Commission. Their efficiency was doubted by both the High Commission and the military commanders, and further differences developed with regard to the organization of the customs regime. It remained to be seen whether the sanctions would have the effect upon Germany that the Allies claimed to expect and for what length of time they would continue in force.

The Spring of 1921.

This proved another troublous period for Germany. There was a practical collapse of the Communist Party in Central Germany during April, and while the agitation in Germany by no means came to an end it was clear that the radical movement in general was steadily waning. Much difficulty over disarmament was experienced with Bavaria, but by the end of May that state had announced its complete agreement with Berlin and its promise to comply fully with the demands of the Allies. Orders for the disbandment of the "Home Guards" were issued, and shortly afterwards Bavaria claimed that it had disarmed below the point of safety. Much apprehension relative to the effects of the customs barrier was expressed and very soon after the system began operation there were loud complaints over its alleged ill effects upon German industry and trade. In May reports that most of Upper Silesia would be awarded to Germany resulted in a serious Polish revolt in that district. The Polish insurgents captured nearly the whole of the industrial area, while the Polish Government disclaimed all responsibility. German "self-defense" organizations, reinforced by contingents from other portions of

Germany, then assumed the offensive. Sharp fighting developed, with the result that the Polish insurgents were defeated and driven back. Negotiations ensued and in June it was agreed that the belligerents should evacuate the plebiscite zone while the Allied troops, which had remained officially neutral, should be placed between the hostile forces. The fighting then ceased.

On April 24 Germany submitted new reparations proposals to the Allies. A total indemnity of 50 billion gold marks was offered and Germany announced her readiness to take an active part in the reconstruction of the devastated regions in France - provided the sanctions were removed. These proposals were rejected by the Allies. Germany then requested the mediation of the United States, pledging herself to pay such amount of reparations "as the President of the United States, after a thorough investigation and examination, may declare to be just and right". President Harding immediately declined to mediate, at the same time expressing the hope that Germany would "quickly submit proposals offering a suitable basis for discussions" and adding that if Germany adopted that course, the American Government would consider calling the attention of the Allied Governments to the matter. Without delay the German Government submitted new proposals to the United States. The German Government was strongly attacked in the Reichstag for these "humiliating" steps, and a Cabinet crisis developed. The Allies announced that the new German proposals were entirely unacceptable. On May 3 the reply of the United States was received at Berlin- the American Government was "unable to consider these proposals as forming a basis for negotiations, acceptable to the Allies". and it urgently advised the German Government to submit direct to the Allies proposals "which in every respect will do justice to its reasonable obligations". Upon the receipt of the American reply the Cabinet resigned, agreeing to remain in office until a new government could be formed. A week later Dr. Wirth of the Centrum Party accepted the post of Chanceller and formed a coalition Cabinet in which five of the outgoing Government were retained.

On May 5 the Allies presented an ultimatum to Germany, calling upon that country to declare its intention of fulfilling without reserve its obligations as defined by the Reparations Commission, to accept unconditionally the guarantees and terms prescribed by that body, to comply without delay with the disarmament demands, and to proceed at once with the trial of war criminals and other unfulfilled portions of the Treaty of Versailles. The penalty for non-fulfillment by May 12 was the occupation of the valley of the Ruhr and "all other military and naval measures that might be required". The next day Germany was informed that the total indemnity had been fixed by the Reparations Commission at 132 billion gold marks, the note containing a detailed schedule of payments. In the meantime the Allies had taken preliminary measures for the occupation of the Ruhr District.

There was considerable difference of opinion in Germany relative to the acceptance or rejection of the ultimatum. The press in general refrained from taking a decided stand in the matter. The Socialists declared their readiness to consider acceptance while the parties of the Right favored rejection.

Those desiring rejection held the opinion that whether or not Germany yielded, the occupation of the Ruhr was but a matter of time. On May 10, by a vote of 220 to 172, the Reichstag approved the recommendation of the new Cabinet to accept the Allied ultimatum without reservations or conditions.

The German Cabinet was not completely formed until sometime later when Dr. Rosen was appointed Foreign Minister and Dr. Rathenau became Minister of Reconstruction. Chancellor Wirth announced his policy as being one of "Conciliation and fulfillment", and the German Government proceeded energetically to plan for compliance with the demands of the Allies. The new Cabinet was considered weak and many claimed that Germany could not fulfill the financial terms of the ultimatum, but none except the radicals of the Right and Left displayed any opposition to the policy of the Government. Arrangements were made for making the first payment of a billion gold marks, the trial of war criminals was begun at an early date and effective disarmament measures were adopted. Public interest now became centered in the Upper Silesian situation, and the month of June ended in a condition of comparative calm. Under all the circumstances the fluctuations of the German mark during the Spring of 1921 were surprisingly slight. Quoted at 62 to the Dollar on April 1, it had declined to 67 by April 22. It then appreciated in value and on May 17 stood at 57. Thence it again declined, reaching 74 on June 31.

The industrial situation had become worse by the end of the period. This was largely because of the coal shortage caused by the troubles in Upper Silesia upon which unoccupied

Germany was mainly dependent for coal. The financial situation remained unfavorable, while trade conditions improved in some respects and grew worse in others. The customs barrier in the Rhineland had a harmful effect, particularly because of the delays in the functioning of the system. Later when the control was somewhat relaxed trade, except as regards the metal and chemical industries, immediately improved.

The food situation was no longer a serious problem. Prices were still high, but there was sufficient food available and the Government announced the early dissolution of the various control agencies.

Labor disputes were frequent and there were a number of strikes. None was of a serious nature except in Munich where a general strike was proclaimed because of the murder of a leader of the Independent Socialists. The effective measures of the authorities soon brought the Munich strike to an end. There were a few demonstrations of unemployed, but, with the exception of Upper Silesia, the country was more free from disorder than it had been for a long period of time. Unemployment continued to be a problem in Germany but it was not so serious as in some other countries of Europe.

There was much disappointment in Germany over the refusal of the United States Government to intervene in the reparations matter, but industrial circles viewed the apparent approach of peace between the two countries with considerable satisfaction. It was felt that peace with the United States would have a beneficial effect upon German trade and, as well, would be to the political benefit of the Reich.

While the inhabitants of the American area were keenly interested in the various problems with which Germany was faced during the spring of 1920, they preserved their habitual calm. Little attention was paid to party politics and the agitation of the separatists was viewed with something akin to indifference. Led by Smeets, who was coming to the fore in the movement, the Rhenish Republican Party, met in Bonn and resolved to increase their propaganda for separation from Germany. The agitation was continued, but it remained relatively unimportant. Even the movement for separation from Prussia became dormant after all parties, except the Communists, had met at Koenigswinter (in the neutral zone) in June and announced that there would be no separation from Prussia, until the occupation of the Rhineland had come to an end. A press rumor that the French Foreign Office favored Dr. Dorten for appointment as German Commissioner for the Occupied Territories-the previous incumbent, Herr von Starck, having resigned[1] -aroused a wave of combined indignation and derision. The report was immediately denied, and the German papers that had published it were admonished.

Living conditions in the American area improved appreciably during the period, and there too the food situation became no longer a serious problem. There were a number of wage disputes and a few small strikes, but on the whole it was a comparatively quiet season in labor circles. Industrial conditions proved much better than had been predicted, despite the customs barrier. The fuel situation was good until June when coal deliveries decreased and caused fears of an actual shortage to arise. Traffic on the Rhine, which had been much interfered with by low water resumed almost its normal volume

[1] Volume I, Chapter 2.

in the latter part of the period when rains caused a rise in the river. Generally similar conditions prevailed in the other occupied zones, and it was a fact that the general economic situation in the Rhineland was better than it had been earlier in the year.

Business and industrial circles in the occupied territories had been profuse in their prophecies that the application of the customs sanctions would cause the economic ruin of Germany, and more particularly of the Rhineland. In this they had been joined by the press of all Germany. The customs barrier proved injurious, but it appeared that the main reason therefor was to be found in the delay caused by the customs examinations; these caused a great congestion at the control stations and a shortage of freight cars developed. After a period of nervousness, business in general became steady, so to speak, and endeavored to accustom itself to the new conditions. Reports soon demonstrated the harmful effects of the customs regime, particularly in the metal, chemical and leather industries. Before long, however, some industries in the American area had secured certain exemptions, and by June the customs control had been relaxed and a number of articles had been put on the free list. These measures were helpful to industry and it soon became clear that the customs sanctions were not having a vital effect upon German trade.

As the customs dues were collected in paper marks and the amounts received was not very large, it became apparent that the financial benefit to the Allies would be relatively unimportant. And as the harmful effect upon German trade and industry was not great, it was seen that as a coercive measure the customs sanctions would not prove of much further effect.[1] Some averred that the ultimate effect upon Allied interests would be detrimental rather than

[1] Their alleged purpose had been achieved when Germany accepted the Allies' ultimatum on May 10.

beneficial. Germans as well as others asserted that the customs sanctions constituted a polictical rather than an economic measure, and before the end of the period discussions relative to the abolition of the sanctions had commenced among the Allies.

Among the prominent visitors to Coblenz in April were the Hon. Hugh Wallace, American Ambassador to France and Marshal d'Esperey of the French Army. During the same month the General Staff of the Belgian Army of Occupation paid a return visit to the American Headquarters and were shown the various activities of the A.F.G. On April 18 a conference of the Chaplains of the Armies of Occupation was held in Coblenz, 32 attending.

In June there came M. Barthou, French Minister of War, M. Loucheur, French Minister for the Liberated Regions, Major General Henry G. Sharpe (formerly Quartermaster General of the United States Army), and the French Generals Pasage, Mazillier, du Racas, du Port and Graziani.

On May 6 General Allen went to Paris for a conference with the American Ambassador. He was again in Paris on May 30 to participate in the Memorial Day exercises at the Suresnes cemetery. Only the ordinary ceremonies were held in Coblenz on May 30, as the American dead had been shipped to the United States some time before and the burial of American soldiers in Germany had been discontinued.

From June 21 to 28 another conference of American military attaches was held in Coblenz. It was attended by the military attaches stationed in Turkey and Egypt as well as by those from European capitals. The Military Intelligence Division of the War Department was represented by Major Marlborough Churchill, General Staff, who had been Director of Military Intelligence during the World War.

Some time before the spring of 1921 General Allen, as Commanding General of the A.F.G., had adopted the policy of turning over to the German authorities those Germans who had committed serious crimes. After the adoption of this policy two Germans murdered an American soldier. Upon their apprehension they were handed over to the German authorities for trial. They were sentenced to death and on June 4, 1921, were guillotined in Cologne.

The Black Troop Question.

The French occupation of Frankfurt and other towns in April 1920 was the signal for a bitter campaign in the German press against the colored troops in the French Army of the Rhine. The propaganda was so extensive and so long continued that there was considerable repercussion in the United States. Many protests against the use of "black troops" in the occupation of "white territory" were made to Washington, and on June 23 the Secretary of State telegraphed General Allen as follows:

> Department recently in receipt of numerous protests based on reports of alleged mistreatment of German women by French colored colonial troops in occupied territories. Please report briefly by cable and at length by dispatch any available information and also your opinion as to truth of alleged outrages.

On July 2, General Allen sent the following letter to the State Department:

> 1. In compliance with your cablegram of June 22, 1920, regarding alleged mistreatment of German women by French colored troops, and in elaboration of my cablegram of June 28th, I submit the following report based on a personal investigation conducted by Colonel Le Vert Coleman, C.A.C., American Liaison Officer with the Commanding General of the Allied Forces of Occupation.
> 2. During the period from January, 1919, to June 1, 1920:
> (a) The average number of negro troops in the French Army of the Rhine was 5,200 men.
> (b) The average number of French colonial troops composed of natives of Africa not of pure negro blood, including distinct races such as Arabs from Algeria, Moroccans, etc., and Negroids,

was 20,000 men. During the entire period from the first day of the occupation in 1918, to the first of June, 1920, 66 cases of alleged rape, attempted rape, sodomy or attempted sodomy have been officially reported to the French military authorities, against their colored colonial troops in the occupied territories of the Rhinelands. Among these cases, there have been 28 convictions, including several cases where the intent was not fully proved but punishment was given by minor courts corresponding to our Summary and Garrison Courts, for indecent proposals and obscene handling of women and girls against their will. There have been 11 acquittals. There have been 23 investigations leading to trials the results of which have not been published yet. There have been 6 cases where the offenders could not be found. The penalties inflicted have been varied; from ten years at hard labor for aggravated cases of rape, to thirty days in prison for indecent mishandling of women.

3. At the present time, the Senegalese Brigade having all left the Rhinelands between June 1st and 6th, 1920, there actually remains but one regiment of troops of Negroid origin, the First Regiment of Chasseurs Malgaches. There are, however, a few individual Negroes or Negroids in the other French colonial regiments.

4. A very violent newspaper campaign attacking the French colonial troops, especially the Negro troops, broke out simultaneously throughout Germany co-incident with the time of the French evacuation of Frankfurt and Darmstadt and has continued up to the present time. It is unquestionably a fact that many gross exaggerations were circulated in the German press concerning the conduct of the French Colonial troops. The allegations in the German press have been, for the most part, so indefinite as to time and place, and circumstance, as to leave it impracticable to verify the alleged facts, or to disprove them.

5. After all proper allowance is made for the natural difficulties which always are to be expected in tracing crimes of this nature, due to the shame and distress of the victims, the great mass of the articles in the German press, by simultaneous appearance all over Germany and by the failure to cite time, place and circumstances sufficiently clear to enable the truth to be ascertained, give to an impartial observer the impression of an adroit political move which would tend to sow antipathy to France in the other lands of the Allied and Associated powers, especially in America where the Negro question is always capable of arousing feeling.

6. The Rheinische Zeitung, and the Koelnische Volkszeitung recently suspended for publishing attacks on the French colored colonial troops, admit under date of June 15, that they employed certain terms and expressions which they might better have omitted due to the imperfection of the news coming for the most part from outside sources, says the Volkszeitung, and from Berlin says the Rheinische Zeitung This tends to bear out the opinion noted above, which is further strengthened by dissentent voices in the South German press which protests against exaggerated accusations by other German papers against colored French troops.

7. These exaggerated attacks in the German press outside of the Rhineland have, in several cases, been refuted by responsible officials (German) and citizens of the Rhinelands.

Herr Kohler, Mayor of Worms, and Herr Bischoff, Police Commissioner of Worms, referring to the Senegalese troops, report to the Inter-Allied Rhineland High Commission, that with the exception of one incident, the Senegalese troops in Worms have not committed any misdemeanor and have been under perfect discipline during their entire stay in Worms.

Herr Levy from Kreuznach, and several Germans have written open letters protesting against what they term unfair exaggerations in the German press against the troops.

8. Among gross exaggerations in the German press may be cited the following:

(a) Claim that there are 40,000 colored French troops in the Palatinate.

(b) Claim that Frankfurt was occupied by 20,000 men entirely formed of Negro (Senegalese) troops. French official report shows that no Senegalese occupied Frankfurt, but first Moroccan and Algerian Tirailleurs and later French troops. (White).

(c) Numerous atrocities in the Saar, where young women are said to have been forcibly abducted, raped, mutilated, killed and their bodies thrown into manure piles. The Burgemeister of Saarbrucken, the inspector of the Caserne Potain, Herr Welsch, proprietor of the manure pile. Wilhelm Roth, caretaker, Herr Geppert, employee, have all given written and oral testimony wholly refuting the accusations.

(d) Claim of the German press that large numbers of young Austrian girls who had come to the vicinity of Mainz to get away from the famine in Austria, were raped. The Austrian Government is reported to have made an investigation through its consular service and to have found that not a single such case had occurred.

(e) Claims in the Nauen Radio Service on April 29, 1920, that the working people of Alsace-Lorraine had protested demanding the removal of the Moroccan Division from Alsace Lorraine, when there was no part of these troops in Alsace-Lorraine.

(f) Investigation by Colonel Bonvialle, commanding the 12th Tirailleurs, May 21, 1920, concerning charges of sodomy near Euskirchen, with medical report indicates that the charges could not be sustained.

9. On the other hand, undoubtedly many instances have occurred where women or girls have been assaulted and some where boys and men have been sodomized by members of the French colored colonial troops. See report above as to the official figures. There are undoubtedly cases which are not included in the official figures due to the natural desire to keep out of obscene notoriety. For example, a case of attempted assault was reported June 14th, 1920, from Saarbrucken which is not included in the French official figures. Some cases will never come to light due to the natural feeling of shame of the women concerned, but they are, in my opinion, cases such as generally occur in any land when soldiery is for a long time quartered upon the population.

10. The impression gained from contact with and observing the French colonial troops is that, as a general rule, they are quiet, orderly and well behaved. Discipline has purely relative value and is hardly of the same order as that which we would require. That the discipline of the Senegalese Brigade was not always good, is established by the incidents which recently occured at Marseilles, when a part of these troops committed serious infractions of discipline when ordered aboard their transport.

11. The attitude of certain classes of German women towards the colored troops has been such as to incite trouble. On account of the very unsettled economic conditions, and for other causes growing out of the world war, prostitution is abnormally engaged in, and many German prostitutes and women of loose character have openly made advances to the colored soldiers, as evidenced by numerous love letters and photographs which are now on file in the official records and which have been sent by German women to colored French soldiers. Several cases have occurred of marriages of German women with French Negro soldiers. One German girl of first-class burgher family, her father a very high city functionary of prominent city in the Rhinelands, recently procured a passport to rejoin her fiance in Marseilles He was a Negro sergeant. Other Negro soldiers have had French wives and the color line

is not regarded either by the French or the Germans as we regard it in America; to keep the white race pure. At Ludwigshafen, when the 7th Tirailleurs left for Frankfurt, patrols had to be sent out to drive away the German women from the barracks, where they were kissing the colored troops through the window gratings.

Conclusions:

1. The wholesale atrocities by French Negro colonial troops alleged in the German press, such as the alleged abductions, followed by rape, mutilation, murder and concealment of the bodies of the victims, are false and intended for political propaganda.

2. A number of cases of rape, attempted rape, sodomy, attempted sodomy, and obscene mishandling of women and girls, have occurred on the part of French Negro colonial troops in the Rhineland. These cases have been occasional and in restricted numbers, not general or widespread. The French military authorities have repressed them severely in most cases and have made a very serious effort to stamp the evil out. The amount of evidence necessary to convict in such cases is very delicate matter to express opinion upon. However, the number of acquittals is not large and there is nothing surprising about these acquittals except in one case where a girl of fourteen years was known carnally. In this case the acquittal followed upon the claim that the girl was consenting.

3. As a rule the number of convictions and the thoroughness of the reports of the investigations and trials indicates the very earnest effort of the French trial authorities to do justice and to stamp out the evil by stern repressive measures. That their sentences are often milder than ours would be, is largely due to extenuating circumstances found in the evidence according to their rules of evidence which are very different from ours, and to the fact that, in general, French courts do not punish these crimes as severely as American and English Courts do.

4. The discipline of the Senegalese Tirailleurs was not always good as evidenced by the refusal of some of them to get aboard transports at Marseilles when ordered to Syria.

This report was given to the American press from which it was copied into European newspapers. German papers published disparagements of the accuracy of General Allen's information, and some alleged that the report showed him to have been unduly influenced "by French propaganda and by his own Francophile feelings". The violence of the "Black Shame" campaign subsided in the late summer, but it was not abandoned, and the propaganda continued in the United States as well as in Germany.

Another press campaign occurred in the spring and early summer of 1921. As before there was much exaggeration concerning the conduct of the colored troops, and in a way the propaganda put forth by the Germans caused a reaction of feeling, even in Germany itself. Some of the German newspapers printed articles stating that the situation was by no means as bad as had been painted, and a number of protests against the campaign were made.

The agitation in 1921 became particularly intense after the publication of a report that three regiments of Negroes had reinforced the French Army of the Rhine. The facts were as set forth in the following extract from a letter written by General Allen to the State Department on April 21:

> The French Rhine Army has recently been reenforced by three regiments of colonial infantry. The reenforcing regiments are the 66th Regiment of Colonial Infantry, (all Moroccans), the 34th Regiment of Colonial Infantry (2 battalions of whites and 1 battalion of Madagascans), and the 42nd Regiment of Colonial Infantry, (1 battalion of whites and 2 battalions of Madagascans). I am officially informed that the three battalions of Madagascans which arrived recently in the Occupied Territories are to tide over the training of the class of 1921, and that they will leave in June. The population tends to confuse the Madagascans with Senegalese negroes and rumors have been circulating that there are three regiments of negroes in the Occupied Territories. The Madagascans who are in the Rhineland are usually known as "Malgaches",

but they are negroes with an infusion of Malay blood. There are also some negroes scattered among the various colonial troops stationed in the Occupied Territories. The 34th and 42nd Regiments of Colonial Infantry are in the same division with a regiment of Senegalese negroes, a fact which may explain why the Madagascan units are identified with negro units by the population. The regiment of Senegalese negroes referred to has not been sent to the Rhineland, but is still stationed in France. The total number of colonial troops, yellow and black, chiefly yellow, now in the Occupied Territories is 27,500.

In August 1920 the German Commissioner for the Occupied Territories had sent to the High Commission a note which made certain allegations concerning the conduct of the colored troops. The following reply was made on April 15, 1921, over the signature of M. Tirard, President of the High Commission:

> I have the honor to send you herein the result of the inquiries made with respect to the memoranda which you sent to the High Commission concerning black troops.
>
> The High Commission first observed that these memoranda only contained 138 accusations during a period which extended over two years, while the strength of the troops against whom the charges were made at certain times, namely during the Armistice, reached a total of about 30,000 men.
>
> On the other hand, out of the 138 accusations, 5 cases concern European French soldiers, 3 cases concern sentinels acting in accordance with their orders, 49 cases were shown to be completely unfounded, 51 were not sufficiently supported by evidence to warrant trial, 30 were considered sufficiently supported to warrant trial (20 of these 30 cases were offenses against morals). Of the cases tried, 13 resulted in conviction, including 4 sentences of imprisonment for terms exceeding five years; 2 cases resulted in acquittal; 7 led to disciplinary punishment; 2 cases were dismissed because of lack of evidence.
>
> While the above figures demonstrate with what rigorous severity the military authorities acted when the complaints were shown to be well-founded, they demonstrate also the incredible carelessness shown in the collection of the documentary evidence, which formed the basis not only of your memoranda but also of the German press campaign

against the black troops.

Further: certain charges were fabricated in every particular by the German officials.

The High Commission decided, by way of a first example, to take punitive measures against the Ludwigshafen official who was responsible for inventing the cases of EMLICH, SCHEY, GREIM, SCHON, and HUGO, as there can be no doubt whatsoever as to these cases since the pretended victims are non-existent.

The High Commission requests you to furnish at the earliest possible date the name of the author of these charges (the author of the report which figures on pages 104-108 of the memorandum of 8th July).

As in 1920, the campaign had subsided by autumn, and only occasional references to the colored troops were made after that time. The American press had published articles on the subject, but there too the agitation died away.

During both years the French press had taken up the cudgels against the German propaganda. An issue was made of the word "black", the French papers asserting that there were no black troops in Germany, and a considerable splitting of hairs developed. The facts were as shown above. There were actually no negro organizations stationed in the Rhineland. But there were some negroid organizations and the other colored units contained numerous individuals of purely negro blood as well as many negroids. It was estimated by Colonel Coleman, who made a thorough investigation of the matter, that in June 1921 from 12 to 15 per cent of the Colonial troops in the French Army of the Rhine were "pure Negroes or approximately pure Negroes".

While it was realized that the German propaganda against the colored troops had over-reached itself and that their conduct had not been as bad as was claimed, the Americans in the

Rhineland were, in general, inclined to sympathize with the feelings of the German population. They preserved a correct attitude in the matter and refrained from open expressions of their views, but they were strongly of the opinion that from all points of view it was unwise to utilize semi-civilized colored troops, whether brown or black, as an occupying force in the territory of a civilized white people. And it was certain that the prescence and conduct of the colored troops in the Rhineland aroused a more intense hatred of the French by the German people.

The Bergdoll Case.

Among the most notorious of those who evaded the draft during the World War was Grover Cleveland Bergdoll, of a wealthy German-American family of Philadelphia He was captured in January 1920 and in the following March was tried by court martial and sentenced to five years' confinement. In May, 1920, having been given permission to visit his home in order to settle some business affairs, he succeeded in escaping, and for months no trace of him was found. His escape created quite a furor in the United States and a scandal was more than hinted at.

In October 1920 the American Headquarters received information that Bergdoll was living in Eberbach, a small town in unoccupied Germany, near Heidelberg in the state of Baden and not far from the perimeter of the Mainz bridgehead.[1] Efforts

[1]He had fled to Canada where he secured a British passport. By way of England and Holland he had reached Eberbach in the vicinity of which his ancestors had lived.

were then instituted to bring about his arrest and delivery to the American military authorities at Coblenz, G-2 (the Intelligence Section) being entrusted with the case as it had occurred in unoccupied Germany.

The matter was first taken up with the German Military Commission[2] which immediately telegraphed the Baden authorities, stating that the American Headquarters desired the arrest and extradition of Bergdoll and his companion, one Stecher, an American of unsavory reputation, who was said to have helped Bergdoll to escape and to be the latter's body guard. The Baden authorities declined to act in the matter on the ground that Bergdoll had never been a soldier and so did not come within the purview of the Armistice Agreement, which provided that deserters or absentees from the Armies of Occupation should be delivered to them by the German authorities. It was further stated that Bergdoll had applied for German citizenship and had said that he had evaded the draft because he did not wish to fight against the country of his ancestors. The War Department was informed of the case and was requested to secure the intervention of the State Department. Having been informed that "the State Department was not inclined to make overtures for the delivery of Bergdoll", the Coblenz Headquarters decided to make no further efforts at the time. The matter was allowed to lie dormant until early in December when work on the case was again begun.

In the meantime disptaches from the War Department had shown it to be greatly interested in the case, and directions

[2] Volume II, Chapter 18.

to make every effort to capture Bergdoll had been received from the Chief of Staff of the Army.

The cooperation of the Secret Service of the French Army of the Rhine was secured and careful plans were made. The general scheme was to entice him to enter the occupied territory, particularly the French area, and there arrest him. As a last resort, an attempt to "shanghai" him would be made; only Germans were to be used in unoccupied Germany, as the Americans entrusted with the case were well aware of the international complications that might ensue should Americans be caught in an attempt forcibly to arrest a person beyond the limits of the occupied territories.

Frequent information of Bergdoll's life in Eberbach was received, the system was working smoothly, and both Americans and French had high hopes of success within a few weeks at the most, when on January 22, 1921, an unfortunate incident occurred and put an end to hopes of the early apprehension of Bergdoll.

Acting under the mistaken authority of their immediate superior officers and contrary to the instructions of the Commanding General, two American operatives of the Provost Marshal's Department (Division of Criminal Investigation), proceeded to Eberbach and there endeavored forcibly to arrest Bergdoll and remove him to the American area. Bergdoll resisted, and one of the operatives fired his pistol, wounding a German girl in the hand. The operatives were then overpowered, one of them being rather badly beaten by the excited crowd, and the two were placed in jail.

As soon as the news was received at Coblenz, General

Allen realized that an international incident had happened. Through his Deputy, Colonel Stone, and the German Commissioner, be informed the German and Baden Governments that the attempt to arrest Bergdoll had been made without his knowledge or authority, expressed his apologies and requested the return of the two operatives to his jurisdiction No reply to this verbal message was received from either the German or the Baden Government.

The incident created some excitement in Baden and caused considerable comment in the German press and in the Reichstag, as well as in newspapers in the United States.

General Allen sent his Judge Advocate to Baden to investigate the case and to request the return of the prisoners. This action brought no results other than information that the operatives would be tried. Another staff officer (a member of G-2) was sent to visit the prisoners at Mosbach whither they had been removed. He reported that they were well taken care of and there seemed no chance of their release before trial.

The Paris edition of the "Chicago Tribune" endeavored to make a sensation of the affair, and succeeded to some extent. It demanded the return of the men, failing which it called on the United States Government to order troops from the A.F.G. to proceed to Baden and forcibly release the prisoners. The paper alleged that the men were suffering from ill treatment and lack of food, and for some time it carried sensational front page articles dealing with the case. The Chicago edition joined in the campaign and with other American newspapers succeeded in inciting many posts of the American Legion to make demands upon the United States Government to bring necessary pressure to bear and cause the early release of the two men.

The unfortunate affair caused much embarassment to General Allen and the War Department. Sensational newspapers and their unthinking readers did not realize that the act of the two operatives was indefensible from an international point of view. Aside from the fact that their action in Eberbach had constituted a technical violation of the Armistice, it was clear that Americans had no more right to make an arrest in unoccupied Germany than Germans had to take a similar step in the United States.

After making arrangements to ensure the fair trial of the two men and to see to their comfort, General Allen continued his endeavors to have them released. He sent a representative to Berlin to put the case before high officials of the Government. They averred their desire to have the two men released, but said that the Central Government was powerless. The Baden Government had complete control of the case and would consider intervention from Berlin as an intrusion upon its prerogatives. Under the Germany system (and more particularly because of the political insecurity of the German Cabinet?) the Central Government could do nothing until the prisoners had been tried. After trial, the officials hoped, arrangements for the release of the men could be made.[1]

[1] It was interesting and enlightening to Americans to find that the political theory of States' Rights was a fact in Germany and that the German States were as jealous of their prerogatives as ever had been Virginia or South Carolina.

It was the impression of the G-2 officer who visited the prisoners in Baden and conferred with the officials there that the Baden Government was "standing on its dignity" and was inclined to use the incident as a means of exploiting its own ends with the Central Government. He was told that the prisoners were under the control of the judicial branch of the Baden Government and that until their trial had been completed, the executive branch could take no steps toward their release.

No active assistance was extended by the American Commissioner in Berlin or the State Department. But General Allen did not let the matter rest. Through G-2 and the American Military Observer in Berlin he continued to bring to the attention of high Government officials his desire for the early release of the two operatives. This method resulted in the receipt of advice as to the best legal defense to be made when the case came to trial and in the practical assurance that the men would be pardoned soon after the trial, which was scheduled to occur early in March.

General Allen decided that he would not take official cognizance of the trial and so would not assign a Judge Advocate to defend the men. Competent legal assistance was to be secured for their defense, but the lawyer would have no official connection with the American Army. At this juncture a member of the Paris Post of the American Legion, a former officer who had served with the Army of Occupation in 1919, arrived in Coblenz. He reported that he had been sent to make arrangements for retaining a lawyer to defend the men, and that sufficient funds for the purpose had been contributed by the American

Legion. He placed himself at the disposal of General Allen and together with G-2 made all necessary arrangements for the defense. The prisoners were visited and furnished with supplies of smoking material, reading matter, etc.

One of the operatives, an ex-soldier, cabled to President-elect Harding that he was "kept mistreated in German jail without help from Coblenz". The matter was referred to General Allen by the War Department, with a request for information as to the status of the case. As a result of the close touch he had maintained with the two men, General Allen was able to show that the complaint was unfounded. He cabled the War Department that both before and after the men had cabled his complaint, he had stated specifically to the G-2 officer who visited him that he had no complaint to make.

The trial of the two men took place at Mosbach on March 21, 1921. A G-2 officer was present as the unofficial representative of General Allen, and several American press correspondents attended the proceedings. The atmosphere of the court room and of the town was distinctly hostile. Bergdoll and Stecher were witnesses and were not restrained from insulting and abusing the two prisoners and Americans in general. The prosecuting attorney made a violent attack on the Armies of Occupation, particularly the A.F.G., and cited the attempted arrest as an example of "the overbearing attitude of the victor". While the presiding judge "hypocritically endeavored to produce an impression of impartiality", he made no attempt to prevent the frequent hostile and insulting remarks of the spectators and witnesses. The counsel for the prisoners made what was characterized as a "brilliant defense", but it was apparent from the beginning that the American operatives would be convicted.

Naef, the civilian operative, who had been in charge of the attempted arrest and had fired his pistol, was sentenced to eighteen months confinement for "unauthorized assumption of authority, coercion and accidental injury". Being a "concurring punishment" the sentence was automatically reduced to fifteen months' confinement. Zimmer, a sergeant, was acquitted of "cooperation" and sentenced to six months' confinement for "assisting in the unauthorized assumption of authority and coercion". Several Germans, who had been employed to assist in the arrest and abduction of Bergdoll and who were tried at the same time, received sentences ranging from six to eleven months' confinement.

While the sentences of the two Americans seemed severe for the offenses cited, it was evident that the influence of higher authority had prevented more serious charges from being preferred.

After having received a report of the trial and the sentences imposed, General Allen, through the same channels as before, brought the matter to the attention of high Berlin officials who assured him that they were doing everything in their power to secure the release of the two Americans. General Allen then referred to their previous statements and inquired as to when he might expect the release of the prisoners. The next day he received the following cablegram from the War Department:

The Secretary of War directs that you bring to a conclusion your negotiations for the release of Zimmer and Naef by making formal application to the proper German authorities for their immediate release. You should impress upon the German Government that the matter is urgent. Report by cable action taken by you and developments as they occur.

General Allen then cabled the War Department that he was dispatching a staff officer to Berlin with a note for the German Government and that he had just received information from the German Commissioner that the early release of the two men was expected.

The above cablegram from the War Department was the first intervention of the United States Government in the case. Both the State and the War Departments had been kept fully informed from the beginning, but beyond indications of their interest neither had taken any action.

On March 28 the following note was delivered in person by a staff officer to the German Under-Secretary for Foreign Affairs, the Minister being out of the city:

Coblenz, 26 March 1921

Sir:

I am directed by my Government to bring to a conclusion my negotiations relative to the American prisoners, Naef and Zimmer, and to request their immediate release.

The delay in releasing these men has made this matter urgent, and my Government therefore directs me to make this formal and earnest application.

I shall be pleased to have acknowledgment of this communication and to be advised of your action at the earliest moment.

Very respectfully

HENRY T. ALLEN
Commanding General
American Forces in Germany.

To
The Minister of Foreign Affairs,
Berlin, Germany.

On March 30, the staff officer returned to Coblenz with the following communication from the German Foreign Office:

Berlin, 29 March 1921.

Dear Major Bagby:

I beg to acknowledge receipt of General Allen's letter of March 26. The present state of the affair is contained in the enclosed Pro-memoria. I trust that the matter will be concluded very soon, and I shall not fail to let you know of the result.

Very truly yours,

E. V. HANIEL.

(enclosure)

FOREIGN OFFICE

A pardon and release of the Americans Naef and Zimmer is exclusively a matter of the Baden Government. The Central Government, however, immediately after the sentence of these two applied to the Baden Government for the pardon of the condemned men. The Prime Minister of Baden also promised to urge their pardon before the Baden Cabinet. The Central Government has today asked the Baden Government to expedite this matter in the above mentioned sense.

Berlin, 29 March 1921.

The foregoing was immediately cabled to the State and War Departments. On March 31 the German Commissioner informed General Allen as follows:

Upon request of the Federal Government the Baden Government has remitted the sentence in the cases of Naef and Zimmer, American Nationals, who had been sentenced to confinement on account of the assault upon Bergdoll at Eberbach, provided their future behavior will be good. Naef and Zimmer will be immediately released from confinement and will return to Coblenz.

The German Commissioner added that the Foreign Office asked that General Allen "kindly award damages" to the girl (Lina Rupp) who had been shot in the hand and that one or more German prisoners confined by the Americans be released. General Allen replied that those matters would be taken under consideration. Full information was immediately cabled to Washington.

As the two operatives had not reached Coblenz by the afternoon of April 1, General Allen requested the German Commissioner to ascertain the reason for the delay. The following telegram, sent by General Allen to the War Department on April 2, shows the next developments:

> German Commissioner hands me telegram from Baden Government showing Naef and Zimmer are not yet released pending receipt of 100,000 marks[1] for girl shot in hand. He intimates German Government powerless in premises. I informed him demand for immediate release was made on Central Government and not on Baden and I would not discuss compensation until prisoners are released. Apparently Bergdoll's money is being used and Baden Government insists on showing independence of Berlin. German Commissioner is impressed with my decision and will give it immediately to Berlin by telephone.

Naef and Zimmer were released at 5 p.m. on April 2 and reached Coblenz early the next morning. With this the incident was closed except with regard to the claim for damages submitted in behalf of Lina Rupp. The sum of 100,000 marks was considered excessive, but it seemed highly probable that the girl (who was 16 years of age) had been deprived of some of her earning capacity. While the A.F.G. had no funds that could be used for the payment of such a claim, General Allen was inclined to forward it to Washington with a recommendation that a reasonable sum be paid her. provided an examination by his medical officers should confirm the German report as to the extent of her injury. The German Commissioner was requested to inform the girl's attorney of the advisability of her coming to Coblenz for such an examination. As the girl declined to come to Coblenz, the facts were reported to Washington and no further action was taken.[2]

The Coblenz Headquarters kept in general touch with Bergdoll after April 1921, but made no active attempt to apprehend him. He continued to reside in Baden, and both French and American operatives were instructed to arrest him should he ever enter the occupied territories. The ill-advised attempt to capture him in January 1921 had spoiled practically every chance of his apprehension by the A.F.G., and it seemed that unless the United States Government could induce the German Government to surrender him, it was highly probable that the notorious slacker would remain at liberty.

The Summer of 1921

While this period opened with a widespread agitation for the abolition of the military and economic sanctions, the actual center of public interest was the question of Upper Silesia.

[2]The matter was finally taken up through diplomatic channels. Early in 1922 a medical officer was sent to Baden to examine the girl and his report was forwarded to the American Embassy at Berlin. He found that the girls earning capacity had been considerably lessened. Several months later General Allen was directed to send all papers in the case to Washington. It was understood that her claim for damages would be submitted to Congress.

The plebiscite district was evacuated by the belligerent German and Polish organizations on July 7. The decision of the Supreme Council, which was to meet at Paris in early August, was eagerly awaited. In the meantime there was a cessation of disorders, but the population in general remained in a state of unrest and the situation could not yet be deemed satisfactory.

The trial of the war criminals at Leipzig was suspended in July after a number of cases had been concluded. Several were sentenced to terms of imprisonment and several others were acquitted. It was announced that no new cases would be tried until the Supreme Council had acted on the verdicts already adopted. Before the suspension of the sittings the French representatives had been recalled and the Belgians had withdrawn their documentary evidence. The French press was unanimous in denouncing the Leipzig trials as "farcical", but the British press was divided in the issue.

The attitude of the Wirth Cabinet evoked favorable comment in the Allied countries. The first billion gold marks had been paid and the Allied Military Control Commission stated that the disarmament was proceeding satisfactorily. There was much discussion anent an extension of the coalition to include representatives of the German People's Party (moderately conservative) in the Cabinet but the plan failed to mature. Much interest was caused by the news that the Government was preparing a new taxation program in order to provide for reparations payments. Reports as to the details of the scheme brought forth protests from many of those who would be affected. There was a slight recurrence of Communist activity. but the end of the period found the party still weaker.

On July 2 President Harding signed the resolution providing for a separate peace with Germany. Negotiations were instituted at Berlin, and on August 25 the treaty was signed there by Mr. Dresel, the American Commissioner, and Dr. Rosen, the German Foreign Minister. The German press expressed the hope that friendly relations would soon be established and pointed with satisfaction to the fact that the question of war guilt had not been mentioned. The treaty read as follows:

Germany

and

The United States of America:

Considering that the United States, acting in conjunction with its co-belligerents, entered into an armistice with Germany on November 11, 1918, in order that a Treaty of Peace might be concluded;

Considering that the Treaty of Versailles was signed on June 28, 1919, and came into force according to the terms of its Article 440, but has not been ratified by the United States;

Considering that the Congress of the United States passed a Joint Resolution, approved by the President July 2. 1921, which reads in part as follows:

> "Resolved by the Senate and House of Representatives of the United States of America in Congress Assembled. That the state of war declared to exist between the Imperial German Government and the United States of America by the joint resolution of Congress approved April 6, 1917, is hereby declared at an end.
>
> "Sec. 2. That in making this declaration, and as a part of it, there are expressly reserved to the United States of America and its nationals any and all rights, privileges, indemnities, reparations, or advantages, together with the right to enforce the same, to which it or they have become entitled under the terms of the armistice signed November 11, 1918, or any extensions or modifications thereof; or which were acquired by or are in the

possession of the United States of America by reason of its participation in the war or to which its nationals have thereby become rightfully entitled; or which, under the Treaty of Versailles, have been stipulated for its or their benefit; or to which it is entitled as one of the principal allied and associated powers; or to which it is entitled by virtue of any Act or Acts of Congress; or otherwise.

. . . .

"Sec. 5. All property of the Imperial German Government, or its successor or successors, and of all German nationals, which was, on April 6, 1917, in or has since that date come into the possession or under control of, or has been the subject of a demand by the United States of America or of any of its officers, agents, or employees, from any source or by any agency whatsoever, and all property of the Imperial and Royal Austro-Hungarian Government, or its successor or successors, and of all Austro-Hungarian nationals which was on December 7, 1917, in or has since that date come into the possession or under control of, or has been the subject of a demand by the United States of America or any of its officers, agents, or employees, from any source or by any agency whatsoever, shall be retained by the United States of America and no disposition thereof made, except as shall have been heretofore or specifically hereafter shall be provided by law until such time as the Imperial German Government and the Imperial and Royal Austro-Hungarian Government, or their successor or successors, shall have respectively made suitable provision for the satisfaction of all claims against said Governments respectively; of all persons, wheresoever domiciled, who owe permanent allegiance to the United States of America and who have suffered, through the acts of the Imperial German Government, or its agents, or the Imperial and Royal Austro-Hungarian Government, or its agents, since July 31, 1914, loss, damage, or injury to their persons of property, directly or indirectly, whether through the ownership of shares of stock in German, Austro-Hungarian, American, or other corporations, or in consequence of hostilities or of any operations of war, or otherwise, and also shall have granted to persons owing permanent allegiance to the United States of America most-favored-nation treatment, whether the same be national or otherwise, in all matters affecting residence, business, profession, trade, navigation, commerce and industrial property rights, and until the Imperial German Government, and the Imperial and Royal Austro-Hungarian Government, or their successor or successors, shall have respectively confirmed to the United States of America all fines, forfeitures, penalties, and seizures imposed or made by the United States of America during the war, whether in respect to the property of the Imperial German Government or German nationals or the Imperial and Royal Austro-Hungarian Government or Austro-Hungarian nationals, and shall have waived any and all pecuniary claims against the United States of America".

Being desirous of restoring the friendly relations existing between the two Nations prior to the outbreak of war:

Have for that purpose appointed their plenipotentiaries:

The President of the German Empire
Dr. Friedrich Rosen, Minister for Foreign Affairs,

and

The President of the United States of America
Ellis Loring Dresel, Commissioner of the United States of America to Germany,

Who, having communicated their full powers, found to be in good and due form, have agreed as follows:

Article I

Germany undertakes to accord to the United States, and the United States shall have and enjoy, all the rights, privileges, indemnities, reparations or advantages specified in the aforesaid Joint Resolution of the Congress of the United States of July 2, 1921, including all the rights and advantages stipulated for the benefit of the United States in the Treaty of Versailles which the United States shall fully enjoy notwithstanding the fact that such Treaty has not been ratified by the United States.

Article II

With a view to defining more particularly the obligations of Germany under the foregoing Article with respect to certain provisions in the Treaty of Versailles, it is understood and agreed between the High Contracting Parties;

(1) That the rights and advantages stipulated in that Treaty for the benefit of the United States, which it is intended the United States shall have and enjoy, are those defined in Section 1, of Part IV, and Parts V, VI, VIII, IX, X, XI, XII, XIV, and XV.

The United States in availing itself of the rights and advantages stipulated in the provisions of that Treaty mentioned in this paragraph will do so in a manner consistent with the rights accorded to Germany under such provisions.

(2) That the United States shall not be bound by the provisions of Part I of that Treaty, nor by any provisions of that Treaty including those mentioned in Paragraph (1) of this Article, which relate to the Covenant of the League of Nations, nor shall the United States be bound by any action taken by the League of Nations, or by the Council or by the Assembly thereof, unless the United States shall expressly give its assent to such action.

(3) That the United States assumes no obligations under or with respect to the provisions of Part II, Part III, Sections 2 to 8 inclusive of Part IV, and Part XIII of that Treaty.

(4) That, while the United States is privileged to participate in the Reparation Commission, according to the terms of Part VIII of that Treaty, and in any other Commission established under the Treaty or under any agreement supplemental thereto, the United States is not bound to participate in any such commission unless it shall elect to do so.

(5) That the periods of time to which reference is made in Article 440 of the Treaty of Versailles shall run, with respect to any act or election on the part of the United States, from the date of the coming into force of the present Treaty.

Article III

The present Treaty shall be ratified in accordance with the constitutional forms of the High Contracting Parties and shall take effect immediately on the exchange of ratifications which shall take place as soon as possible in Berlin.

In witness whereof, the respective plenipotentiaries have signed this Treaty and have hereunto affixed their seals.

Done in duplicate in Berlin this twenty-fifth day of August, 1921.

Rosen
Ellis Loring Dresel

The murder of Mathias Erzberger[1] on August 26 caused a political crisis that for a time occupied the center of the stage. The entire German press expressed indignation and fear of grave political developments. Great excitement ensued. The parties of the Left and a large portion of the Centrum declared the crime to have been a political murder, and their newspapers engaged in a violent attack against "the Bolshevists of the Right", some predicting a new revolution. So serious was the situation that the President decreed the suspension of publications the contents of which tended to incite the overthrow of the Government and of meetings and demonstrations should there be apprehension that the debates might be of the same tenor. Several papers of the Right were immediately suspended, and the Government instituted a campaign against the monarchists in general. Throughout Germany, there were demonstrations of protest against the murder and the "ceaseless agitation and unscrupulous methods of the Right",

[1]Erzberger was a member of the Centrum Party and a stormy petrel of German politics. He introduced the famous peace resolution in 1917 and after the revolution, as the Chief of the German plenipotentiaries, signed the Armistice Agreement. He became Minister of Finance in 1919 and in 1920 was forced to resign because of disclosures relative to his conduct of that office. Because of the general antipathy against him, he did not hold high office afterwards. However, taking up with the Socialists, he remained a power in German political life.

that in Berlin bringing together the largest assemblage since the revolution of 1918. The repressive measures of the Government were well received except in Bavaria, where much resentment was expressed and relations with Berlin became strained. Much opposition was expressed by papers of the Right in various parts of Germany, but the Presidential decree was enforced, and even in Bavaria several papers were suspended. A revival of open class hatred developed. In a number of places mobs entered public and private buildings and destroyed everything they could find that related to the old order of things[1]. Heated discussions in the press continued for some time, but no serious disorders developed, and by the end of September the political atmosphere was again comparatively quiet.

The tense relations between the National and Bavarian Governments was one of the outstanding events of domestic politics during the summer. The resentment caused by the Presidential decree mentioned above was greatly increased by demands from Berlin for the raising of the state of siege which had long been in force in Bavaria. The Bavarians took the stand that the control of the press and of meetings and demonstrations was a matter for Bavaria alone - that Bavaria should be governed from Munich and not from Berlin. The same principle applied to the preservation of order, and the question of the continuance or ending of the state of siege could be decided by the Bavarian Government without intervention

[1]Instances of this kind occurred in the French bridgeheads of Mainz.

from Berlin. In 1919 parts of Bavaria, including Munich, had been ruled by a so-called Soviet government, and that state did not care to run chances of a second experience of the kind. But very recently there had been a great Socialist and Communist demonstration in Munich; under the pretext of protesting against the high cost of living, the assemblages had advocated the overthrow of the existing government. Incited by the steadily increasing cost of living, the population was "gradually becoming alarmed", and Bavaria could not dispense with martial law. Negotiations between Bavaria and Berlin were instituted. The results of these negotiations, which were in the nature of a compromise, were rejected by the Bavarian Cabinet which resigned after the Assembly and disapproved the rejection. The new Cabinet, composed largely of members of the old but with a new Premier, took a more moderate view of the controversy with Berlin, accepted the compromise and promised to raise the state of siege. By the end of the period relations with the National Government were normal once more.

The Supreme Council met at Paris on August 8.[1] The Germans were tensely interested in the outcome, particularly with regard to Upper Silesia, and great was their disappointment when, unable to reach a decision upon Upper Silesia, the Supreme Council referred the question to the League of Nations. Both the Polish and the German elements were disappointed over the delay in reaching a decision and complained of "terrorism"

[1]The United States looked upon the Upper Silesian question as a strictly European affair and did not participate officially in the meetings.

by armed bands, but leaders of both sides exhorted their partisans to preserve order until the final decision was reached. The population in general, seemingly weary of disturbances, remained passive. The Allied reinforcements, the dispatch of which had been decided upon some time before, arrived in the plebiscite district, but no occasion for their intervention arose. The decision of the League of Nations was eagerly awaited, but up to the end of September practically no outward manifestations of excitement had occurred.

During the summer of 1921 economic conditions in Germany were better than at any time since the revolution of 1918. Towards the end of the summer food prices commenced to rise after having decreased appreciably earlier in the season, but it remained true that the average family had been better off with regard to the necessities of life than for long before. In July the effects of the troubles in Upper Silesia had caused a coal shortage; by the end of September the shortage had been relieved. While there were many strikes and lockouts, the number of unemployed was far less than during the spring months. Industrial activity increased throughout Germany, the depreciation of the mark (from 75 to the dollar on July 1 to 117 on September 30) greatly stimulated the export trade, and a number of ships were added to the German merchant marine. The Wiesbaden Agreement between France and Germany, providing for reparations deliveries in kind, was signed in August and caused satisfaction in Germany. In September an economic agreement with Italy was concluded; this promised to prove of much benefit to German trade. Many fears of an impending financial collapse were expressed and Germany's economic problem remained difficult and far from solution. But in the final analysis the conditions of life in the summer of 1921 were considerably easier for the average German than they had been since the Allied and Associated armies first set foot upon German soil.

Generally similar economic conditions prevailed in the occupied territories, although the cost of living was higher there, particularly in the American area, than in unoccupied Germany. But there were many less strikes in the occupied territories and the population was not subjected to the agitation and excitement that occurred from time to time in the interior. Low water in the Rhine caused a congestion in shipping, despite the ending of the tug-boat strike in early July. This factor, together with the diverting of Ruhr coal to unoccupied Germany in order to make up the deficit arising from the decreased deliveries from Upper Silesia, brought about a coal shortage in the Rhineland. But the Allies reduced the monthly reparations deliveries by approximately 30% and the Upper Silesian mines increased their output, the abolition of the economic sanctions resulted in a renewed activity among the industries and the grain and potato crops proved very good. All things considered the Rhinelander's conditions of life were at least as good as those in unoccupied Germany; and for him, too, the summer of 1921 was, by comparison with those of 1919 and 1920, a prosperous season.

At its August meeting the Supreme Council had decided to maintain the military sanctions but to abolish the economic measures on September 15. The latter action was to be dependent upon the payment on August 31 of one billion gold marks by Germany and upon that country's agreement to the establishment of an Allied license control system in the Rhineland. The German press voiced considerable objection to these provisos and to the failure to provide for the withdrawal of the troops from Duesseldorf, Duisburg and Ruhrort. It had been hoped that the sanctions would be entirely removed, and much disappointment was expressed. At the same time industrial circles contemplated the early removal of the customs barrier with unconcealed satisfaction. Germany paid the required one billion gold marks, but as the Government failed to agree to the license control system until late in September, the customs sanctions were not removed until September 30. The additional French and Belgian troops, brought up in the spring, then commenced to withdraw to their respective countries.

While the United States did not participate in the application of the sanctions, General Allen, in his capacity as American Representative with the High Commission, was frequently called upon in the meetings of that body to give his opinions in the matter. He believed that as the sanctions were alleged to have been adopted as coercive measures they should have been abolished as soon as their purported object had been achieved. In May the Germans had accepted the Allied ultimatum, and therefore there was no longer any reason for maintaining the sanctions. He claimed also that the economic sanctions were harmful rather than beneficial to Allied interests as they lessened Germany's ability to pay the reparations demands.[1] It transpired that the British and Belgian High Commissioners agreed with General Allen's views, while the French High Commissioner adhered to the opinion that the sanctions should be continued in force.

The Allied license control system did not commence operations during the period. The High Commission had been entrusted with its organization, but could not come to an agreement. General Allen believed that it might prove injurious to American trade. He succeeded in having an American member placed on the committee assigned to organize and conduct the system, and he took effective steps to prevent discrimination against American goods.

The population of the American area remained orderly throughout the period, and there were no outward manifestations of excitement over the various events in unoccupied Germany. There was an infiltration of Communist influence, but it proved of no practical importance. The local Socialist papers took part in the campaign against the reactionaries that occurred after the murder of Erzberger, and heated discussions took place between them and papers of the opposite persuasion. The local political parties were not active until towards the end of the period.

There was a revival of separatist activity in the southern portion of the Rhineland and several meetings were held in other

[1]In fact it seemed that the occupation of the Rhineland was also harmful rather than beneficial to the reparations interests of the Allies, as it too lessened Germany's ability to pay.

localities, but the movement made practically no progress. Dr. Dorten came to Coblenz to attend a meeting and was called before the American military authorities. He was directed to abstain from any speeches that would tend to cause excitement and so interfere with the preservation of order. He promised to comply with these instructions and took advantage of the opportunity to declare himself as favoring the separation of the Rhineland from Prussia but not from Germany. He was informed that the Americans were not concerned with the internal politics of Germany except in so far as they might affect the maintenance of order, and it was reiterated that the American authorities would not permit a change in the local government by other than constitutional means. Dorten did not attend the scheduled meeting which was attended by a very small number of people, and he left the area on the following day.

In September a tremendous explosion occurred in the chemical works at Oppau, near Ludwigshafen in the French zone. Over 400 were killed, the total casualties approaching 1000 in number. Help was immediately extended by the French Army of the Rhine, and expressions of sympathy were made by the High Commission and the various military commanders.

July 4 was celebrated throughout the area, the program being similar to that of 1920. The French national holiday on July 14 was marked by the usual salutes and the decoration with the fourragere of the 2nd Battalion of the 6th Field Artillery and Companies A, B, and D of the 1st Engineers - a graceful act of the French. The Belgian holiday was celebrated on July 21 with ceremonies similar to those of the previous year.

In July a delegation of Spanish officers visited the area, and several days later a number of French naval officers, headed by Admiral Ratye of the Superior Naval War College, came to Coblenz and were extended the usual courtesies. Other visitors during the summer were General Degoutte; Lord Hardinge, the British Ambassador to France; Senators McKinley of Illinois, Walsh of Montana and Robinson of Arkansas; Representatives Brooks, Barkley and Montague; General Oskar Paul Enckel, Chief of Staff of the Army of Finland; M. Loucheur, French Minister of the Liberated Regions; and the former French Premiers Rene Viviani and Georges Leygues.

In July a committee of the Chamber of Commerce of the United States, the members of which were studying business and political conditions in Europe, visited Coblenz at the invitation of General Allen, remaining for two days. The members, all prominent in the American business world, were as follows:

Mr. Joseph H. Defrees, President of the Chamber of Commerce of the U.S.;
Mr. John H. Fahey of Boston;
Mr. R. F. Lamont of Chicago;
Mr. E. A. Filene of Boston;
Mr. Silas H. Strawn of Chicago;
Mr. John J. O'Connor of Washington.

The delegation displayed a keen interest in conditons in the Rhineland, the policies of the American authorities and the relations of the latter with the Allied and German authorities. Under the instructions of General Allen the members of his staff, of both the A.F.G. and the High Commission, as well as he himself, discussed matters very frankly with the delegation and endeavored to furnish it with an accurate picture of actual conditions. The delegation was greatly interested in the policies and intentions of the Allied countries and of Germany as they appeared to the Americans on the Rhine, in the probable effects of these policies upon European peace and American trade, and in the effect upon all countries concerned of a withdrawal of the A.F.G. and the American representation with the High Commission. The delegation departed for Berlin with expressions of appreciation and of a belief that the return of normal conditions in Europe and in American trade therewith would be much delayed by an early withdrawal of the Americans from occupied Germany. After returning to the United States the delegation conferred with the Secretary of State and published a lengthy report in which the views mentioned above were embodied.

A company of infantry was sent to Flirey, France, in August to attend the dedication of a monument to the American Expeditionary Forces, presented by the French Government.

The maneuvers of the "A.F.G. Division" were held from September 23 to 30. General Pershing was present on several days, and at the conclusion of the maneuvers he inspected and reviewed all combat troops and their horse and motor transportation. The command was assembled on the Weissenthurm Flying Field, and it proved the last occasion upon which all troops of the A.F.G. were assembled together.

General Pershing, now Chief of Staff as well as General of the Armies of the United States, spent a week as the guest of General Allen, during which he made a thorough inspection of the varied activities of the A.F.G.

In July General Allen proceeded to London for a conference with the Hon. George Harvey, American Ambassador to

England. General Allen attended the meeting of the Supreme Council in August, carrying with him special reports, drawn up by his staff, on matters bearing upon the Upper Silesian question and the sanctions. His position at the Supreme Council was that of special adviser to Mr. Harvey (whose status was that of an observer for the United States), but he participated in a number of conferences held outside of the meetings of the Supreme Council and gave his considered views on the subjects within his special field. During the middle of September General Allen made a cruise of several days in the Baltic on the American battleship "Utah", visiting Copenhagen enroute to board the vessel. Debarking at Danzig, he proceeded to Berlin where he conferred with the American Commissioner and several members of the German Government.

The Last Quarter of 1921.

The leading domestic issues were now the proposed extension of the coalition cabinets, both of the Nation and of Prussia, and the troublous question of Upper Silesia. In both cabinets only three parties - Majority Socialist, Centrum and Democratic - were represented. The existence of each cabinet was very insecure, and it was desired to strengthen each by the addition of members of other parties. All efforts thus to broaden the basis of the National Cabinet failed, but before the end of the period success was achieved in the Prussian Cabinet, and with four parties represented it became relatively much stronger.

The decision of the Council of the League of Nations, in the matter of Upper Silesia was delivered to Germany on October 20. It provided for a partition of the industrial district, the most important and valuable portion of which was awarded to Poland. A "Mixed Upper Silesian Commission" was to guarantee the economic unity of the district and to adjust economic matters there during a period of transition which was to last fifteen years, and an agreement to this effect was to be concluded between Germany and Poland. There were violent protests throughout Germany, demonstrations were held at Berlin and other large cities, and the situation in Upper Silesia again became tense. As a result of the decision, the Wirth Cabinet resigned on October 22. The political parties were unable to agree upon a basis for a new coalition in the Government, and Dr. Wirth was again entrusted with the formation of a Cabinet. The Democrats consented to the retention of Dr. Gessler, Minister of National Defense, a Democrat, but otherwise would not participate unless the German People's Party were represented in the Cabinet. This left only the Centrum and the Majority Socialists represented in the new Government; as the Cabinet was not supported by a majority it was called an "Administrative Cabinet". While the new Cabinet entered upon its duties on October 26, several vacancies were left unfilled, particularly that of Foreign Minister, and it had not been completely formed by the end of the year. Chancellor Wirth again declared his policy to be one of "fulfillment".

There was much talk of a rejection of the decision relative to Upper Silesia. October 22 was celebrated as a national day of mourning, although the Socialist parties did not participate, and public amusements were suspended for 24 hours.

The German Government protested that the awarding of Upper Silesian territory to Poland violated the principles of self-determination, but in view of the threats of the Allies of adopting "such measures as might be deemed necessary to insure the provisions of the Geneva verdict", the decision was accepted. Upper Silesia remained comparatively quiet, although complaints of "terrorism" continued to be made by both sides. Negotiations between the Germans and Poles were instituted, and by the end of the period it was announced that an economic agreement had been reached.

Ex-King Ludwig of Bavaria died in Hungary in October. His funeral took place at Munich early in November and was the occasion for a huge royalist demonstration There were no disturbances in Bavaria, as the working classes maintained an attitude of reserve. In other parts of Germany the Socialists staged demonstrations in opposition to the royalist agitation as well as in celebration of the third anniversary of the revolution of 1918. They announced that they were on the alert with regard to reactionary tendencies and there was much discussion in the press of the Right and Left, but no untoward incident occurred.

After the lifting of the state of siege in Bavaria on October 15, friction with the Central Government ceased, to be renewed before the end of the month when the National Minister of Justice announced that his department would inspect a certain prison in Bavaria. The Reichstag passed a bill at that effect, but soon revoked it because of protests from Bavaria. The friction was lessened towards the end of the period, but as the question of Bavaria's prerogatives remained a live issue, tense relations with the Central Government might develop at any time.

The failure of Ex-Emperor Carl to regain the throne of Hungary by his attempted coup was welcomed by all parties in Germany. Even the monarchists were opposed to a restoration of the Hapsburgs, as they feared that it would greatly endanger the ultimate fusion of Austria with Germany.

There was a revival of Communist activity in November when about 100 Communists, who were serving prison sentences for participation in the March disorders, went on a hunger strike. The Communist Party attempted to exploit the incident and were alleged to have been responsible for the unemployed riots that occurred in November. The Government refused to release the hunger strikers and the riots were sharply repressed by the police. The agitation continued, however, and in December, pending further investigation of their sentences, a large number of participants in the March "uprisings" were released.

An interesting incident of the period was the trial of persons implicated in the Kapp Putsch, which took place at Leipzig in December. Herr von Jagow was sentenced to five years' detention in a fortress, while the proceedings against other accused persons were suspended. Ludendorff testified during the trials, and the Socialist press demanded that charges be preferred against him also. Nothing came of this, however. In the meantime Kapp remained in Sweden, and other men who had been prominent in the revolt were ostensibly in hiding.

The treaty of peace between the United States and Germany was ratified in October and on November 11, Armistice Day, the ratifications were formally exchanged in Berlin. Normal diplomatic relations were now resumed: the American Commissioner, Mr. Dresel, became Charge d'Affaires of the Embassy in Berlin and a similar official was sent to Washington to take up his duties in the German Embassy.

In December the Allied Military Control Commission announced that the dismantling of all unauthorized fortifications in unoccupied Germany had been practically completed and that plans had been made for the destruction of the remaining fortifications in the occupied territories.[1] At about the same time Chancellor Wirth claimed that all "self-defense" organizations had been disbanded.

A number of state and municipal elections took place during the period. The results in most cases showed appreciable gains by the parties of the Right.

The reparations question was the most important issue during the last three months of 1921. The Loucheur-Rathenau, or Wiesbaden, agreement for the partial payment of reparations by deliveries in kind was signed early in October. The total value of German deliveries to France up to May 1, 1926, was not to exceed seven billion gold marks. This amount was to be credited to the reparations account, but not more than one billion was to be credited in any one year. Supplementary agreements fixed the amounts of railway rolling stock, machinery and livestock which Germany was required to deliver to France, determined the prices for coal deliveries, and granted Germany the right to export coal if she fulfilled the reparations demands.

After the decision of the Council of the League of Nations

[1] Volume II, Chapter 13.

with regard to the partition of Upper Silesia had been made public, German financiers predicted a financial collapse and claimed that the fulfillment of the reparations demands was now impossible. Bills providing for the new taxation program were introduced into the Reichstag, bringing forth a storm of criticism from the members. The Government declared its willingness to carry out all reparations agreements until the Allies recognized that Germany could not fulfill the entire demands. Industrial leaders offered to grant a loan to the Government provided the national railways were converted into private enterprises; this was strongly opposed by the Socialists and the railway employees. Up to the end of the year no particular progress in the taxation program or in the "credit action" of German industries had been made. Both Rathenau and Hugo Stinnes, the "wizard of German industry", endeavored to raise a loan in England, but failed to achieve success.

The Reparations Commission visited Berlin and announced a belief that Germany could meet the next reparations payments which were due in January and February 1922, and demanded that every effort to do so be made. Chancellor Wirth replied that as the German Government had not been able to raise the necessary loans either at home or abroad, he could not guarantee payment of the amounts due in January and February. He then stated that his Government was obliged to ask for a moratorium. The discussions with the Reparations commission were continued, and at the end of the year the issue remained unsettled.

A conference between Lloyd George and the French Premier Briand was held at London in December. It was decided to hold a meeting at Cannes in early January, for the particular purpose of discussing the reparations question and methods of stabilizing the economic situation of Europe.

At its meeting in August the Supreme Council had provided for a commission of financial and military delegates to examine into the costs of occupation and such reductions as in their opinion might be effected. The Commission met in Paris during October. The American delegates were General Allen; his Finance Officer, Lieutenant Colonel E. A. Sturges; Mr. R. W. Boyden, United States Unofficial Representative with the Reparations Commission; and Colonel R. H. Hess, U. S. Army Reserve, American Finance Representative.

The question of limiting the total costs of occupation to 240 million gold marks per annum, in accordance with the agreement of June 16, 1919, was discussed. The American delegation stated that it was without authority to agree to any limitation upon the costs of the American occupation short of the actual costs to the United States of its occupying forces; it believed that the costs of occupation should be reduced by reducing the number of troops in the Rhinelands; and General Allen proposed to cooperate by reducing the number of American troops. The British delegation took the stand that the purpose of the meeting was to determine how the yearly sum of 240 million gold marks was to be divided among the Governments concerned, and if any Government exceeded the amount allotted it, the excess cost should be borne by that Government. The French delegation objected to any reduction of troops that would interfere with French policy in the Rhineland, and questioned the competency of the Commission to discuss the matter. General Weygand, Chief of Staff to Marshal Foch, stated that he would have to withdraw from the meeting if a "reduction in effectives was discussed". The Belgian delegation supported the French stand, with the result that the discussions were confined to considering the costs of occupation without limitation.

An estimate of the costs of the American troops, based on 8500 effectives (after May 1, 1922) was approved by the Commission. This estimate fixed the direct American costs at $14,492,000 direct costs and the indirect costs (supplies and services in kind) at approximately 57 million paper marks. General Allen announced that he would make every effort to keep the American costs within the amounts stated in the estimate, and even to reduce the costs below these figures.[1] The total number of the troops of occupation had been fixed at 132,000. Immediately after General Allen proposed to reduce the American troops to 8,500, the other delegations announced their intention of increasing their forces by the number of American troops withdrawn and of estimating their costs on that basis. The British were opposed to this action, but stated that "in self-defense" they had to participate.

The action of the Interallied Military Control Commission in demanding the partial destruction of certain plants belonging to the Deutsche Werke, a great corporation whose factories had formerly produced munitions of war, caused heated protests in Germany. It developed into a political issue when the Reichstag passed a unanimous resolution of protest. The Government requested revocation or modification of the order and also protested,

[1]For a statement of the costs of the American Occupation and German payments thereof, see Volume 1. Chapter 9.

against a demand that the manufacture of certain types of Diesel motors be prevented. Both matters were discussed by the Conference of Ambassadors, but no definite decision was arrived at.

One of the outstanding features of the economic situation was the depreciation, accompanied by violent fluctuations, of the mark. On October 1 the American dollar would buy 122 marks, and on October 17 it was worth 191 marks. Four days later the dollar was quoted at 161 marks, and on October 31 its value was 180 marks. Then came the great November depreciation and the violent fluctuations. The tabulation set forth below shows the course of the mark during the last two months of 1921 -- the figures being the exchange value of the dollar in German marks.

November 2	194
" 5	250
" 7	350
" 9	246
" 12	287
" 15	260
" 26	296
" 30	240
December 1	192
" 5	230
" 14	177
" 16	198
" 19	178
" 21	171
" 31	184

The great depreciation of the mark was said to have been due to the belief of foreign financial circles that Germany's bankruptcy was at hand. The partial recovery between November 7 and 9 was caused by strong rumors that a moratorium would be granted to Germany. The fluctuations in general were attributed to the alternating pessimistic and optimistic rumors with which the financial atmosphere was alive at that time. Business in general was greatly stimulated and there was a boom on the stock exchanges in Germany. The fluctuations occasionally caused mild panics, however, and in the beginning of December a number of banks failed because of speculation, while "thousands of speculators were ruined". Gambling in foreign currency had become so widespread that the Reichstag passed a law to curb it.

There was another rush of foreign buyers to Germany and prices rose sharply. Every portion of Germany was affected by the influx of foreigners, and certain industries in neighboring countries were partially paralyzed by the sudden cessation of orders. The regulations against large sales to foreigners did not prove entirely effective but they served to stop the rush. In many parts of Germany anywhere from 25% to 100% extra was charged foreigners by retail merchants, restaurants and hotels; in the occupied territory this was not permitted in the in the case of members of the occupying forces and the High Commission or their families

Prices rose rapidly as the mark depreciated in value, and in November there was a large increase in the number of wage disputes, strikes and lockouts. The labor unions were very active, but in general employers were disposed to grant reasonable increases of wages. The cost of living continued to increase and demands for higher wages kept pace with it throughout Germany. Wage disputes and strikes grew somewhat less in December but the rise in the exchange value of the mark did not result in a decrease of the cost of living. At the end of the year a general strike of postal employees was threatened, while a railroad strike had been called on December 29.[1]

Food conditions were comparatively good as regarded the amount available, but the prices had risen greatly. Due to the long period of low water in the rivers and the unusually cold weather, which had caused many canals to freeze, a coal shortage developed and became a serious economic problem. Railroads had to curtail their traffic and many industries were adversely effected. At the same time the number of unemployed was far less at the end of December than it had been at the beginning of the period.

The end of the year found the conditions of life, particularly of the salaried people and those with fixed incomes, considerably worse than they had been during the summer; for them, in fact, the conditions were more adverse than at almost any time since 1918. In this connection the index figures of the cost of living, in terms of the 1913 cost which is assumed to be 100, are of interest:

January 1920	854
January 1921	944
September 1921	1062
December 1921	1550

A glance at the financial situation at the end of 1921 will be of interest. During December the expenditures of the German Government had exceeded the receipts by over 19 billion paper marks. The floating debt was almost 247 billion paper marks and the number of Reichsbank notes in circulation amounted to 117 billion paper marks. At the beginning of the year the floating debt had been 153 billion paper marks and the Reichsbank notes in circulation 80.8 billion paper marks.

The end of 1921 found Germany faced with many difficult problems

[1] Volume I (Chapter 2) and Volume II (Chapter 14).

and possessed of grave apprehensions as to her future. But her people were hard at work, "the chimneys were smoking" throughout the country, and by comparision with other nations unemployment had ceased to be a problem.

The population of the occupied territories retained a keen interest in the various questions of the time and the conditions of life continued to be generally the same as in unoccupied Germany. As usual, however, the occupied territories, and more particularly the American area, experienced many less strikes. Wage disputes were constant, but most of them were settled without the occurrence of strikes, and the only serious one was the railroad strike called on December 29, and this was settled in short order.

Great satisfaction was expressed over the abolition of the economic sanctions, and much dissatisfaction was voiced when it was reported that France had decided to maintain the military sanctions. As the industries in the American area were already working nearly to capacity, the effect of the removal of the customs barrier was not very great, and it was thus shown that the economic sanctions had not been nearly so harmful as was claimed. At the same time it should be remembered that as licenses were still required for the passage of goods, trade continued to be hampered.

With the exception of the separatists, the politicians in the Rhineland were not very active during the period. The Dorten group met at Cologne in November, but the attendance was very small and Dorten indicated his belief that the movement had failed.

The Smeets group now became more active. (Their organ, the "Rheinische Republik", published in Cologne, had been suspended for two weeks by the German police, acting under the general authority of the Presidential decree issued in August). Early in December they held a meeting at Bonn (French area). The meeting was orderly, but several days afterward Smeets was arrested by the German police. The High Commission ordered him to be released pending investigation.[1] This action caused great indignation, and all political parties except the Independent Socialists and the Communists denounced Smeets and affirmed their loyalty to Prussia and the nation. In the middle of December all parties except the two mentioned above met at Koenigswinter in the neutral zone. A resolution was passed denouncing all separatist propaganda and emphasizing the loyalty of the Rhinelanders to the German nation. The formation of a Rhenish state within Germany should take place only in accordance with the German Constitution, and it was insisted that no vote on this subject should be taken during the period of the occupation. While the Independent Socialists did not participate in the meeting, they issued a notice declaring themselves to hold the same views. Other meetings held in the occupied territories showed clearly that the separatist movement was opposed by all but a small minority of the German population. Smeets claimed to be optimistic, but it was apparent that there was no chance of separatism succeeding unless it was forcibly imposed on the Rhineland.

Early in October a number of German ex-officers who were holding a meeting in Crefeld (Belgian area) were arrested by the Belgian military authorities. Sixteen of the thirty-two placed on trial were found guilty of having made speeches prejudicial to the safety of the troops of occupation, or of having weapons in their possession, or of having entered the occupied territories without proper passports.

[1] Volume I, Chapter 2.

The sentences imposed ranged from one to six months' imprisonment and from 1000 to 10,000 marks fine. The arrests and convictions resulted in the expression of much indignation in the German press.

The coming into force of the treaty of peace between the United States and Germany raised an interesting question as to the status of the American troops and of the American zone. Both had been governed by the Armistice Agreement except in so far as General Allen had permitted the Ordinances of the High Commission to apply in the American Area. Early in November he submitted the following to the State Department:

> Shall the American Forces in Germany continue to be governed by the powers granted the Commanding General in the Armistice Agreement, as at present?
>
> or
>
> Shall these forces be governed as provided in the Rhineland Agreement, annexed to the Treaty of Versailles, contemplated by Article 1 of the new Treaty?

As a decision was not received up to the end of the year, General Allen pursued the same policies as before, and the ratification of the Treaty of Berlin caused no change in the conduct of the affairs in the American area.

A reduction of the American troops having been ordered by the War Department, some 3600 troops left for the United States during the period and one brigade was disbanded on December 31, 1921.

The reduction in strength caused a recurrence of rumors that all American troops were to be withdrqwn at an early date. The German population of the area were quick to express their desire that some American troops remain until the end of the occupation, and the German press contained articles along similar lines. The German Government made no official statement, but high officials asked that Washington be unofficially informed that Germany hoped that the United States would not give up the American area. The High Commissioners and the Allied military commanders in the Rhineland also expressed their desire that the United States continue to be represented in the occupation. Frequent press dispatches from the United States announced the early withdrawal of the American troops, but no indication of its intentions in the matter came from the United States Government.

A composite battalion of picked men was present at Paris in early October when General Pershing placed the Medal of Honor upon the tomb of the Unknown Soldier of France. The same battalion was sent to London to participate in the ceremonies incident to the decoration of the British Unknown Warrior by General Pershing. In the middle of the month General Allen, accompanied by a detachment of non-commissioned officers, proceeded to Chalons-sur-Marne; representing the United States he there received the American Unknown Soldier and escorted the remains to Le Havre where he handed them over to the Navy. The Unknown Soldier was placed on board the U.S.S. "Olympia" and transported to Washington for burial in the National Cemetery at Arlington.

November 11, Armistice Day, was celebrated in the American area with appropriate ceremonies.

A rather severe epidemic of influenza and pneumonia occurred in the American area in December. Both the troops and the civil population were affected and there were a number of deaths.[1]

[1] Volume II, Chapters 5 and 26.

Christmas was again characterized by gifts to poor German children. A different system was employed: Each organization assembled the poor children in its immediate vicinity, furnished a Christmas tree and amusements, and distributed presents to the little guests. As before, these kindly actions evoked appreciative comments from the German people and press.

General Allen was present at the decoration of the French Unknown Soldier, but was unable to participate in the ceremonies in London because of his attendance at the meeting in Paris to discuss the costs of occupation. In early November he went again to Paris to attend a similar meeting, and late in the month he visited London for a conference with the American Ambassador.

Among the prominent visitors to Coblenz during the period were Marshal Petain and General Degoutte of the French Army and Major General H.L. Rogers, Quartermaster General of the United States Army.

CHAPTER NO. 2.

The Interallied Rhineland High Commission.

Organization; Powers and Duties; Initial Legislation; In the American Area; Liaison with German Authorities; Activities; Appendices.

Reference

General ---------------------- -A.M.S. Volumes I and IV.

CHAPTER NO. 2.

THE INTERALLIED RHINELAND HIGH COMMISSION.

page

ACTIVITIES (con't)

APPENDICES

THE INTER-ALLIED RHINELAND HIGH COMMISSION.

ORGANIZATION.

The Rhineland Agreement, which formed an annex to the Treaty of Versailles and was signed with it on June 28, 1919, provided for the establishment of the Inter-Allied Rhineland High Commission[1]. The inception of the idea of such a commission

[1] A.M.G., Volume IV, Page 1381.

and the discussions that resulted in the Rhineland Agreement are shown in previous volumes of this work.[2]

Article 2 of the Rhineland Agreement provided that -

> There shall be constituted a civilian body styled the Interallied Rhineland High Commission and hereinafter called the High Commission, which, except insofar as the Treaty may otherwise provide, shall be the supreme representative of the Allied and Associated Powers within the Occupied Territory. It shall consist of four members, representing Belgium, France, Great Britain and the United States.

Since May 1919 the Inter-Allied Rhineland Commission[3] had been the agent of the Supreme Economic Council in the Occupied territories. Between the signing of the Treaty and its ratification in January 1920, this Commission had been largely engaged in making preparations for the coming into power of its successor, the High Commission.[1] The initial organization of the High Commission was essentially the same as that of its predecessor.

The High Commission consisted of four High Commissioners, appointed respectively by the Governments of the United States, Belgium, Great Britain and France, the appointment of each being made with the approval of the other Governments. The French High Commissioner was President. Each High Commissioner was assisted by a Deputy High Commissioner and by technical advisers selected by the former. Two Secretaries-General, one English speaking and the other French speaking, were appointed by the High Commission and were to be present at all its meetings. The seat of the High

[2] A.M.G., Volume I, (Page 450,) and Volume IV, (Page 1378.)
[3] A.M.G., Volume I, Chapter 18, Page 457.

[1] The ordinances and instructions published on January 10, 1920, by the High Commission had been prepared by the Rhineland Commission in conjunction with the military authorities.

Commission was at Coblenz where one or more meetings were held each week.

Each High Commissioner was present, in person or by representative, at all meetings. Decisions were adopted by majority vote, each High Commissioner having one vote In case of a tie, the President could give a casting vote. Should a High Commissioner be dissatisfied with a decision he might appeal to his Government, but in urgent cases such an appeal would not operate to delay the execution of the decision, which was carried out upon the responsibility of those who voted for it. The results of the meetings were recorded in minutes which were communicated to those concerned.

Usually questions which came before the High Commission had been studied in advance by one or more of the permanent Committees, the members of which were selected from among the technical advisers of the High Commissioners. Each committee was composed of four members, one from the department of each High Commissioner. The committee system facilitated the work of the High Commission, as in the majority of cases the opinion of a committee was adopted by the High Commissioners with but little discussion. The permanent committees were -

Administrative
Coal
Communications
Economic
Financial
Intelligence
Legal
Requisitions.

Further details regarding the organization and procedure of the High Commission will be found in Appendices 1 and 2 to this chapter.

The High Commissioners.

As originally constituted the High Commission consisted of the following High Commissioners and Deputy High Commissioners:

Belgium: Baron Rolin-Jaequemyns
M. Fernand Cattoir

France: M. Paul Tirard
M. Amedee Rousselier

Great Britain: Sir Harold Stuart
Mr. Melcolm Arnold Robertson.

The American Representative.

Before the ratification of the Treaty of Versailles on January 10, 1920, the representative of the United States (Mr. Pierrepont B. Noyes) took an active part in the preparations for the establishment of the High Commission and in the drafting of the initial ordinances and instructions. At that time his powers and status were the same as those of the other officials who were soon to be named High Commissioners. But the failure of the United States to ratify the Treaty caused a change in his status after January 10. He could not be designated as High Commissioner and therefore could not be an official member of the High Commission. It appeared that until the United States were officially represented on the High Commission, the latter could not be a legally constituted body.[1] The legal question was ignored, and the problem as to the participation of the American representative

[1] The German Government might well have raised this point; that it refrained from doing so was probably because of a belief that an expression of opinion by the American representative would have the same weight as the votes of the High Commissioners, and possibly because of a desire that the military government of the occupied territories should come to an end as quickly as possible.

was solved by an agreement that Mr Noyes, with the approval of the State Department, should continue to represent the United States, but that his presence at the meetings of the High Commission should be unofficial and he should have no vote. His technical status having been settled, the American Representative (as he came to be designated) proceeded to organize his department, the functions of which became similar to those of the other departments. He attended all meetings of the High Commission and, while technically he did not vote, his opinions had practically as much influence as if they had been votes.

In order to keep in close touch with the activities of the High Commission, the American Commanding General assigned an officer of his staff (Colonel D. L. Stone, G. S. as liaison officer and military adviser to the American Representative.

The German Commissioner.

While the Rhineland agreement was still in embryo, the German Government requested permission to appoint a German official to co-operate with the High Commission and to constitute the channel through which the population of the occupied territories might address the High Commission. This German official would be the supreme representative of the German Republic and of the federated states concerned (Prussia, Bavaria, Hesse, Baden and Birkenfeld), and would alone be authorized to negotiate with the High Commission. This request was repeated in a note date July 11, 1919,[2] and with certain reservations was granted by the Allied and

[2] A.M.G. Volume IV, Appendix No. 51 (Document III, Page 124)

Associated Powers on July 29, 1919.[1] The German Government was given to understand that the appointment of such an official was not provided for in the text of the Rhineland Agreement; that the individual chosen for the position must previously be approved by the Allied and Associated Governments, who might recall their approval at any time; and that his competence could extend only to those matters which, in accordance with the provisions of the German Constitution, came under the authority of the central German Government. The right of the High Commission to enter into relations with any local German authority whatsoever regarding matters within the latter's competence, was also reserved.[2]

The German official was designated the Reichskommissar, and became generally known as the German Commissioner.

The German Peace Delegation, in view of the fact that the occupied portion of Germany was governed by the laws of the National Government and of five federated states as well,[3] requested that before publishing its ordinances the High Commission consult the German Commissioner and secure his opinions thereon. The Allies agreed that such a procedure might be useful, but stipulated that there would be no obligation upon the High Commission to follow it.[4a] During 1920 and 1921 the High Commission several times consulted the German Commissioner before publishing ordinances, but it was by no means a general rule. In fact, during 1921 his opinion was seldom ascertained before the publication of an ordinance and then only because of his protest.

[1] A.M G., Volume IV, Appendix No. 51 (Document III, Page 135)
[2] Ibid - Page 136.
[3] Ibid - Page 125.
[4] Ibid - Page 136.

Herr von Starck was the first German Commissioner, serving as such until the summer of 1921.

POWERS AND DUTIES.

Upon the declaration of a state of peace between the Allied countries and Germany, the High Commission became the supreme representative of the Allies in occupied Germany. Its authority extended throughout the occupied territories, which included the four bridgeheads (Cologne, Coblenz, Mainz and Kehl) as well as all German territory on the left (west) bank of the Rhine. Technically it had no power in the American area, but practically it had, by virtue of a modus vivendi which had been agreed upon by the American Commanding General and the High Commission.[1]

The High Commission was created to secure the maintenance, safety and requirements of the forces of occupation, and it was given the power to issue such ordinances as might be necessary to fulfill its mission. But the civil administration was to remain in the hands of the German authorities and under the authority of the central German Government, except in so far as it might be necessary for the High Commission to adapt that administration to the needs and circumstances of military occupation.

The power to declare a state of siege (martial law) in the occupied territories or any part thereof was given to the High Commission. Upon such declaration the military authorities would assume control, but they were to act in combination with the High

[1]For a description of the way in which the difficult situation with regard to the American area was met, see A.M.G, Volumes I (Chapter 18) and IV (Page 1376 et seq.)

Commission and with its approval in the issue of decrees and proclamations or in the matter of intervention in the civil administration.

The Rhineland Agreement provided that the ordinances of the High Commission should have the force of law; these ordinances related to criminal and civil jurisdiction and pertaining matters and to administrative matters affecting the forces of occupation. The power of adapting the civil administration to the needs and circumstances of military occupation was exercised by the suspension of modification of such German laws as might prejudice the maintenance, safety or requirements of the forces of occupation. This power extended to all German laws published after the ratification of the Treaty of Versailles; before these could become effective in the occupied territories, they had to be submitted to the High Commission for examination and decision as to their admissibility.

Under the system provided for in the Rhineland Agreement three varieties of laws were effective in the occupied territories-

> Ordinances of the High Commission, most of which applied to the occupying forces as well as to the civil population.
>
> German laws, which applied to foreigners who were not attached to the armies or the High Commission, as well as to Germans.
>
> Laws of the countries participating in the occupation, which governed their respective nationals officially present in the Rhineland.

The ordinances were transmitted to the military authorities through the Headquarters of the Allied Armies of Occupation and to those German authorities who were charged with their enforcement. In so far as Germans were concerned, the ordinances were considered as an integral part of German law applying to the occupied territories. German courts, in applying or interpreting German law, had to take the ordinances into consideration. As a matter of fact, the ordinances took precedence over German laws, as all provisions of the latter that conflicted with the ordinances were automatically suspended.

INITIAL LEGISLATION.

On January 10, 1920, the High Commission published the following ordinances and instructions:[1]

> Ordinance No. 1 - Legislative power of the High Commission, orders of the military authorities and operation of German laws and regulations in the Occupied Territories.
>
> Ordinance No. 2 - Criminal and civil jurisdiction and offences relating to the occupation.
>
> Ordinance No. 3 - Movement of persons; postal telegraphic and telephonic communication; restrictions imposed on the press and public meetings; possession of and trade in arms and ammunition.
>
> Ordinance No. 4 - Extension of the ordinances of the High Commission to the Kehl Bridgehead.
>
> Ordinance No. 5 - Procedure to be followed in the settlement of industrial disputes.
>
> Ordinance No. 6 - Powers and duties of the Interallied Rhineland Railway Commission.
>
> Instruction No. 1 - Use of the Allied forces in Occupied Territories for maintenance of public order.
>
> Instruction No. 2 - Duties of German authorities in matters of security and police.
>
> Instruction No. 3 - Notification of appointment of German officials.
>
> Instruction No. 4 - Notification of disease.
>
> Instruction No. 5 - Exemption from the jurisdiction of courts.
>
> Instruction No. 6 - Supervision and inspection of prisons.
>
> Instruction No. 7 - Permits to carry arms and ammunition.

In large part this legislation was the result of the experience

[1]Ordinances and Instruction will be found in Appendix 2 to this Chapter.

gained by the forces of occupation during the Armistice, and in modified form it embodied many orders issued by the military authorities. It was felt that as peace had been declared, many of the restrictions that had obtained since the beginning of the occupation should be removed. But the provisions of the ordinances concerning the movement of individuals, telephone and telegraph communication, the settlement of industrial disputes, duties of German officials, inspection of prisons, and the issue of permits to carry arms and ammuntion - all these were in keeping with the policies followed by the armies during the period of the Armistice. It was the endeavor of the High Commission, however, in so far as was compatible with the interests of the occupation, to restore its former liberty to the population and to charge the German authorities with the responsibility for the maintenance of order and the operation of public utilities. Some of the more liberal provisions of the ordinances are set forth below.

Germans traveling from one portion of the occupied territories to another or entering or leaving unoccupied Germany, required no special authority, the only requirement being that every-individual above the age of 14 should have an identity card.

Newspapers and books could be published freely and without censorship before publication. The High Commission reserved the right to forbid or suspend the publication or sale of any periodical or book of a nature prejudicial to public order or to the security or dignity of the High Commission or the forces of occupation.

There was no restriction upon public meetings other than that 48 hours notice of all political meetings had to be given. Any meeting that might endanger public order could be forbidden or dissolved.

In principle there was to be no censorship of the mail, telephone or telegraph, but the High Commission from time to time caused the correspondence of certain suspicious persons to be censored.

Non-interference with the German administration was the policy of the High Commission. However, should a German official disobey its ordinances or the orders of the military authorities, he was removed from office and sometimes expelled from the occupied territories. And if an official appointed by the German authorities proved to be undesirable, his appointment was rescinded at the demand of the High Commission.

As regarded German criminal jurisdiction, the personnel of the forces of occupation and of the High Commission were not subject to it. The military personnel could be punished only by military tribunals, while members and employees of the High Commission could not be tried by any court in the Rhineland without the consent of the High Commission.

In civil suits, members of the military forces or personnel of the High Commission, in their private capacity, might be cited to appear before German courts, but if the findings of the court were considered unjust, such persons could appeal to the High Commission.

IN THE AMERICAN AREA.

The failure of the United States to ratify the Treaty of Versailles caused a difficult situation to arise in the American area after January 10, 1920.[1] The problem was simplified by the decision of the American Commanding General to make the ordinances effective in the American area by their publication in orders by his Headquarters. By making a few minor reservations, he enchanced the prestige and authority of the High Commission without relinquishing any of the powers inherent in him under the Armistice Agreement and the laws of war. With a few exceptions, all American military orders that conflicted with the ordinances were revoked by Civil Affairs Bulletin No. 58, issued on January 31, 1920.[2] The reservations and supplementary provisions made by the American Headquarters are briefly described below.

Article 16 of Ordinance No. 3 required that notice of political meetings must be given 48 hours in advance of the meeting. Experience had demonstrated that industrial (labor union) meetings had been more prolific of trouble than had political meetings; hence in the American zone a 48 hours advance notice of both political and industrial meetings continued to be required.

Article 8 of Ordinance No. 3 provided that -

> Any person who wishes to take up his residence in the Occupied Territories shall make application in writing to the German authorities of the locality in which he wishes to reside. Such authorities may give the required permission and shall, within three days, communicate the decision reached to the representative of the High Commission in the Kreis concerned.

1 A.M.G., Volume IV, Page 1376.
2 Volume I, Chapter 3 (Appendix No. 1.)

Coblenz, the seat of the High Commission as well as of the American Headquarters, was very much congested. Therefore the American military authorities continued to require that those who desired to remain in Coblenz for more than 20 days would have to secure the consent of the American Headquarters (through the German authorities).

A technical state of war continued to exist between the United States and Germany. Hence, until the two countries should be at peace, the German courts would not be permitted to exercise civil jurisdiction over members of the American forces (as was provided by Article 15 of Ordinance No. 2).

As the Ordinances contained no provision regarding such matters, the American authorities reiterated that German physicians were prohibited from treating or giving medical advice to members of the American forces suffering from venereal disease.

Article 29 of Ordinance No. 2 forbade persons trading with the public to sell to any member of the forces of occupation, or to any Allied official, any article at a higher price than that usually paid by the German public. With a view to the enforcement of this provision, the American military authorities required price lists to be posted in restaurants and cafes and articles for sale in shops to be marked with their respective prices.

Subject to the above changes and supplementary provisions, the initial ordinances were permitted to be effective in the American area as well as in the remainder of the occupied territories. With very few exceptions all ordinances and instructions issued by the High Commission between January 10, 1920 and the end of 1921 were permitted to apply in the American zone. Article 2 of Ordinance 49 provides that -

> The High Commission alone shall have the right to decide the suitability of the premises intended for the residence of its members, officials and employees, with respect to their rank and office and of their functions or duties.

The American Commander could not admit that in the matter of billets in the American area the High Commission could have right superior to his own; in addition, the congested conditions in Coblenz required special regulations. It was agreed, therefore, that the article cited should not be effective in the Landkreis or Stadtkreis Coblenz.

LIAISON WITH GERMAN AUTHORITIES.

The German note of July 12, 1919, regarding the interpretation of the Rhineland Agreement, assumed that after the ratification of the Treaty, administrative or supervisory officials would no longer be attached to the German authorities as had been the case during the Armistice.[1] This was admitted in the reply of the Allied and Associated Powers, which stated that the High Commission would maintain fixed representatives charged with the duty of securing liaison between the local German authorities, the local military authorities and the High Commission itself.[2] Accordingly, the High Commission appointed a representative in each Kreis or corresponding administrative district in the occupied territories. The Kreis representatives were to act as liaison officers between the military authorities and the local German

[1] A.M.G., Volume IV, Appendix No. 51, (Document III, Page 131)
[2] Ibid - Page 142.

authorities. In addition, it was their duty to transmit the orders of the High Commission. While in no way was it intended that they should supervise the local German administrations, many attempts to extend their authority were made. The duties of the Kreis representatives were clearly defined in the ordinances and instructions of the High Commission.[3]

ACTIVITIES.

During the early months of 1920 the High Commission was engaged in solving numerous administrative problems which had arisen as a result of its having taken over many of the functions formerly exercised by the military authorities. The critical political situation in unoccupied Germany and the accompanying events caused the High Commission to take action in several instances.

The Kapp Putsch.[1]

The coup d'etat, generally known as the Kapp Putsch, occurred in Berlin in the middle of March 1920 and was followed by widespread disorder in unoccupied Germany. The great industrial region in the valley of the Ruhr River was one of the centers of disorder. As the Ruhr District adjoined the British and Belgian zones of occupation, it was feared that the disorder might spread to the occupied territories. As a measure of prevention the High Commission adopted Ordinance No. 12, which empowered its local representatives to prohibit all meetings or gatherings of a nature that might endanger the security of the forces of occupation, and to direct the population to keep off the streets at night. This

[3] And see Appendix 3 to this Chapter.
[1] Volume I, Chapter I.

ordinance was applied for short periods of time in several districts in both the British and Belgian areas.

On March 26 the High Commission adopted Ordinance No. 16, permitting censorship over the mail, telephone and telegraph to be imposed for a period of one month. A censorship was immediately imposed in the Belgian area where radical agitation was being conducted.

While maintaining a neutral attitude toward the dissensions in unoccupied Germany, the forces of occupation would permit no disorder or provocative demonstrations in the occupied territories. As a result of this stand and of the preventive measures of the High Commission, there was no disturbance of any consequence in the Rhineland during the eventful spring of 1920.

The Ruhr Revolt.[1]

The collapse of the Kapp government was followed by a serious uprising in the Ruhr District, which was generally termed the Ruhr Revolt. Much hard fighting ensued, and the struggles between the revolting workmen and the troops of the Reichswehr[2] involved developments in the occupied territories that called for the intervention of the authorities of occupation.

On March 19, after a combat with the "Red Troops", some 1500 men of the Reichswehr sought refuge in the British area. These troops were disarmed and interned by the British. After negotiations

[1] Volume I, Chapter I.

[2] Literally, the National Defense Organization - the newly adopted designation of the combined German Army and Navy. Later the German Army was designated as the Reichsheer, or National Army.

covering several weeks, and with the consent of the Allied military authorities, these troops were permitted to return to unoccupied Germany.

Another problem arose when, on April 6, several thousand Red troops took refuge in the British zone after having been defeated and pursued by the Reichswehr: they declared their belief that the Reichswehr would execute them without trial if they remained in the Ruhr District The fugitives were disarmed and interned. The guarding and feeding of so many persons for a prolonged period was impracticable; hence the High Commission proposed to the Allied Governments that they negotiate with the German Government for the return, under suitable guarantees, to unoccupied Germany of the interned fugitives. An amnesty having been proclaimed by the German Government, practically all of the Reds had been returned to unoccupied Germany by April 21.

After this there were no further difficulties in the British or Belgian zones. The American zone remained unaffected during the entire period, except that the floating population was increased by the arrival of a number of undesirables who were promptly deported.

The Occupation of Frankfurt and Darmstadt.[1]

During the operations which resulted in the quelling of the Ruhr Revolt, the German Government sent more troops into the neutral zone than were authorized under the existing agreements governing the number of effectives therein. The Germans claimed

[1]Volume I, Chapter 1.

that the number of troops used were the minimum necessary to the restoration of order in the Ruhr District. Their contention was not admitted, and, after much discussion pro and con, the French Government directed that the cities of Frankfurt on the Main and Darmstadt be occupied by the French Army of the Rhine. Upon the receipt of the instructions from his Government, the French Commanding General requested the High Commission to declare a state of siege in the bridgehead of Mainz. He claimed that such action was necessary in order to ensure the security of his troops and their communications. The High Commission complied with the request and a state of siege, limited to the Mainz bridgehead, was proclaimed on April 6 - the first instance of the kind that had occurred since the coming into force of the Treaty of Versailles.

The High Commission requested that it be informed as to the non-military measures which had been put into effect in the Mainz area as a consequence of the state of siege. The French Commander replied that the state of siege was of a restricted nature and that the non-military orders issued by him prescribed only such measures as were indispensable to the security of his troops. They included the supervision of German officials and public utilities by the military authorities, the prohibition of strikes and of unauthorized meetings and demonstrations, the establishment of military courts for the trial of civilians, and the suspension of newspapers for a few days. Later the restrictions were made less severe, and on May 2, upon the recommendation of the French Commander, the state of siege was ended by the High Commission.

With the ending of the state of siege in the Mainz area, the last difficulties which had arisen from the upheavel caused by the Kapp *Putsch* were settled in so far as the High Commission was concerned.

The Relief of Mr. Noyes.

In May 1920, Mr. Pierrepont B. Noyes, the American Representative, was informed by the State Department that as the Treaty of Versailles had not been ratified by the United States, it was considered advisable that the American Civilian representative be withdrawn. At the close of the meeting of the High Commission on June 2, Mr. Noyes informed the High Commission of his imminent departure for the United States and of his regrets upon ending his pleasant association with them. The President, speaking for the High Commission, stated that Mr. Noyes had furnished the inspiration for the Rhineland Agreement and that he had taken a preponderant part in the work which had culminated in the establishment of the High Commission. The President laid stress upon the disinterestedness, good faith and devotion displayed by Mr. Noyes and expressed the keen regret held by the High Commission upon his departure.

Appointment of General Allen.

Major General Henry T. Allen, the Commander of the American Forces in Germany, was informed on May 21 that, at the request of the State Department, the Secretary of War had agreed that upon the relief of Mr Noyes, General Allen could replace him. This responsibility was to be in addition to his duties as Commanding General. Accordingly, upon the departure of Mr. Noyes, General Allen became the American Representative with the High Commission.

Realizing that his military duties would render it impossible for him to devote as much time to High Commission matters as could the High Commissioners, General Allen, in order to keep in daily touch with the High Commission, appointed Colonel David L. Stone, G.S., as his delegate. The latter's position was practically that of a Deputy High Commissioner; he attended all meetings and represented the United States in the settling of all but the more important questions.

General Allen kept his functions as American Representative entirely separate from those as Commanding General, and in his relations with the High Commission he acted, in so far as was practicable, as if he had been actually a civilian representative of the United States. Similarly, although he was a member of General Allen's Military Staff, Colonel Stone conducted himself in his work with the High Commission as a civilian rather than as a military official.

The Arrest of Dr. Dorten.

Dr. Dorten had been the leading German figure in the separatist movement in the Rhineland, and in May 1919, with the support of the French military authorities, had attempted to proclaim a Rhenish Republic, with himself as President. His attempt, such as it was, came to grief because of the refusal of General Liggett, Commanding the Third American Army to permit any interference with the Constituted German authorities except through due process of German law. In June Dorten made an unsuccessful attempt to oust the German authorities at Wiesbaden. Separatist agitation continued thereafter, but the vast majority of Rhinelanders were strongly opposed to it, and the movement failed to gain headway.[1]

Dr. Dorten maintained his residence in Wiesbaden (French area) and there, on July 24, 1920, he was arrested and removed to unoccupied Germany by armed police from Frankfurt.[1] In response to an inquiry, the Wiesbaden police authorities informed the High Commission that Dorten had been arrested in accordance with an order issued by the Frankfurt Chief of Police. The latter based his authority for the arrest upon a warrant, issued on June 11, 1919, which accused Dr. Dorten of high treason within the meaning of Articles 81 and 82 of the German penal code.[2]

During the peace negotiations, the Allied and Associated Powers had decided to protect those persons in the occupied territories who, during the Armistice period might have violated German political, financial or commercial laws, without having acted in such a way as to require prosecution by the authorities of occupation. The German Delegation protested, proposing to substitute an amnesty which would include the individuals convicted by Allied or Associated military courts as well as those who had violated German laws. The proposal was rejected, and in accordance with the decision mentioned, Article 31 of Ordinance No. 2 was adopted -

> Without the consent of the High Commission, no judicial proceedings shall be instituted or continued and no punitive measures shall be taken against any person in the Occupied Territories for any administrative or political act done during the period of the Armistice.

[1]A.M.G, Volume I, Chapter 15. Page 377.
[1]The French had withdrawn from Frankfurt on May 17.
[2]Upon several occasions during the Armistice period, the German authorities had endeavored to arrest Dr. Dorten, but had been prevented by the military authorities of occupation who acted under the powers conferred upon them by the Armistice Agreement.

Dr. Dorten had been arrested on a warrant issued during the Armistice and because of political acts committed during that period. Therefore the High Commission considered that the action of the German police constituted an offense against its authority and that of the French military authorities; as the arrest had not been made with the consent of the High Commission, it was a violation of Article 31 of Ordinance No. 2 and of Ordinance No. 27 as well. The offense was an aggravated one in that it had been committed by German police from unoccupied Germany and that Dr. Dorten had been removed from the occupied territories.

While the French High Commissioner was the only one who displayed interest in Dr. Dorten and his political aspirations, the other members of the High Commission felt that they could not permit the German authorities to violate its ordinances with impunity. A demand for the immediate return of Dr. Dorten to Wiesbaden, in order that he might have opportunity to present his defense as provided by Ordinance No. 27, was made upon the German Commissioner. At the same time the High Commission informed the German Commissioner that pending the results of an investigation as to the implication of the Regierungspraesident and Chief of Police of Wiesbaden, both officials were suspended from office until further notice. The High Commission also informed the Commanding General of the French Army of the Rhine that it had no objection to the opening of a legal investigation of all persons suspected of complicity, especially the Regierungspraesident and Chief of Police of Wiesbaden.

The investigation made by the French military authorities disclosed the fact that the arrest of Dr. Dorten had been ordered by the Regierungspraesident of Cassel, in unoccupied Germany, and had been made under the direct instructions of the Chief of Police of Frankfurt. Several days before the arrest there had been two separate attempts to arrest Dorten; the Frankfurt police were in Wiesbaden when those attempts were made; Dorten had been shadowed by means of information furnished by the police of Wiesbaden, who had communicated his movements by telephone; these facts were sufficient, in the opinion of the High Commission, to implicate the Regierungspraesident and Chief of Police of Wiesbaden and to justify its action in suspending them from office.

The German Government immediately released Dr. Dorten, admitted that the execution of the order for his arrest had constituted a violation of the ordinances of the High Commission, and expressed its regrets therefor. It was further stated that the Prussian Government had instituted an inquiry into the circumstances of the arrest, and assurances were given that those found culpable would be punished.

On September 23, the German Government submitted the report of its investigation. The Chief of the Intelligence Bureau for the Province of Hesse-Cassel, whose station was at Cassel, had been found to have been responsible for the arrest of Dr. Dorten. It was in consequence of information received by telephone from the Cassel Intelligence Bureau, that the Chief of Police at Frankfurt had come to the erroneous conclusion that the projected arrest of Dorten had been approved by the high administrative authorities. The Chief of the Intelligence Bureau at Cassel was removed from office, and the Chief of Police at Frankfurt was severely reprimanded.

Upon the receipt of the report of the German Government, the Dorten incident was considered to be closed. At a later date as it appeared that personally he had not been involved, the High Commission reinstated the Regierungspraesident of Wiesbaden.

Arrest of German War Criminals.

During the Armistice, and even after the coming into force of the Treaty of Peace, a number of Germans charged with offenses committed during the war had been arrested by the Belgian and French military authorities of occupation. Some of these had been tried by military courts in the occupied territories and sentenced to fines and imprisonment, while others had been taken into Belgium and France and there brought to trial. On January 24, 1920, the German Commissioner protested against this procedure and requested the High Commission to bring about the immediate release of the persons arrested. His contention was that in arresting persons charged with offenses committed during the war, the countries participating in the occupation were acting without legal authority; the arrest in German territory of German nationals for offenses committed abroad was a matter for the German authorities only, and where the arrest of an alleged "war criminal" was desired, formal application therefor should be made to the German authorities. He contended further that there was no authority for the removal of such persons from German territory for trial in a foreign country.

In reply to a number of similar protests, the High Commission, on March 25, 1920, informed the German Commissioner that the matter of arrests made during the period of the Armistice was not within its competence, and that the question of those made after the coming of peace would be referred to the governments concerned. The High Commission then gave instructions that no further arrests should be made, but as the military courts were disinclined to abandon their jurisdiction, many difficulties attended the release of those already arrested.

Numerous requests for permission to arrest alleged offenders residing in the American area had been submitted to the American Headquarters by the Belgian and French authorities. These requests had always been denied, the Americans holding that as the Allies had agreed that, for the time being, alleged war criminals should be tried by the German courts at Leipzig, Article 228 of the Treaty of Versailles no longer applied. On January 22, 1920, through an error of judgment on the part of a subordinate American officer, the American military police arrested a Dr. Hermann of Neuwied (American area) and delivered him to the Belgian military authorities. The arrest had been made at the request of the Belgians, who promptly removed the prisoner to Tourani. As soon as the error was discovered, the Americans requested the release of the prisoner. It developed that he had been tried "by default" before his arrest, and that he had already entered upon the service of his sentence of imprisonment; his delivery to the Belgians was then shown to have been doubly in error, as the American policy was against the arrest of persons sentenced "by default", whether or not they were war criminals.

Upon receipt of a protest from the German Commissioner the High Commission submitted the question of Dr. Hermann's arrest to the Belgian Government. The request for his release was repeated from time to time by the American Headquarters, but months passed before the matter was settled. Finally on August 13, the High Commission was informed that the Belgian final court of appeal had set aside the verdict of the court of appeal at Brussells. The warrant of arrest was cancelled, and Dr. Hermann was released forthwith.

French Attitude towards Civilian Control.

The High Commission, in June 1920, authorized certain railway officials to carry arms. Somewhat later, General Payot, the Director General of the Communications and Supply of the Allied Armies, suspended this order. Through the Commanding General of the Allied Armies of Occupation, the High Commission informed General Payot that the cancellation or suspension of an order of the High Commission was not within his power.

This incident was indicative of the attitude of the French, civilian as well as military, towards civilian control of the occupied territories. Upon several occasions the French military authorities encroached upon matters within the jurisdiction of the High Commission; having placed a *fait accompli* before the High Commission, the French apparently expected it to approve the action taken. Such action was invariably defended by the French High Commissioner, which made it very difficult, and often impossible to have the High Commission take action in the way of a reprimand or caution to military commanders who had exceeded their authority.

There were very few instances of this kind in the Belgian area; whenever the Belgian military authorities displayed a tendency to exceed their authority, the Belgian High Commissioner was the first to call their attention to the limitations upon their powers. Neither the British nor the American military authorities ever came into conflict with the High Commission because of having acted in excess of their authority.

Relief of Sir Harold Stuart.

Sir Harold Stuart, the British High Commissioner tendered his resignation in October 1920. He had taken an important part in the organization of the High Commission and in the drafting of its ordinances, and he was as noted for his frankness and impartiality as he was for his intellectual powers and wide administrative experience. His departure involved a great loss to the High Commission and gave rise to expressions of regret from all sides. It had developed that, without pre-arrangement of any kind, his views and those of the American Representative had coincided upon almost every question of policy that had come before the High Commission.

Mr. Malcolm Arnold Robertson, the Deputy High Commissioner was then appointed British High Commissioner, being succeeded in the former position by Lieutenant Colonel Rupert S Ryan.

French Time in the Rhineland.

Ordinarily there is a difference of one hour between French (Western European) time and the Central European time used in Germany. The adoption of summer time in France thus caused the same time to be used in both countries between March and October. In September 1920 the Headquarters of the Allied Armies of Occupation requested the High Commission to direct that Western European time be used in the Rhineland during the winter of 1920-21.[1]

[1]During the winter of 1919-20, Western European time was used on the railroads in the occupied territories and, except in the British area, as civil time also. Watches were advanced an hour as soon as trains entered unoccupied Germany, and, when it was ten o'clock in the British area it was nine o'clock in other parts of the Rhineland. The following winter civil time and German time were the same throughout the occupied territories.

This was objected to by the American and British forces who took the stand that it might prove disadvantageous to the industrial welfare of the occupied territories because and an additional hour of darkness during the working day would be entailed, and that in view of the shortage of coal, a change of time was highly undesirable. The American Representative pointed out that the request for a change of time was based upon the probable confusion that would arise in the event of mobilization, with Western European time used in France and Belgium while Central European time prevailed in the Rhineland; this confusion could be avoided by the issue of a military order providing that in mobilization orders the times prescribed should read one hour earlier in the occupied territories, and the economic and industrial life of occupied Germany should not be injuriously affected because of the remote contingency of mobilization. The request was then modified so as to affect railroad time only. To this the American, British and Belgian Commanders objected, stating that civil time and railroad time should be the same. Mr. Robertson, the British High Commissioner, stated at first that under no circumstances would he agree to a change of time; he was then informed by the French High Commissioner that the Allied Military Committee at Versailles considered it necessary that the railroad time should coincide with French time, whereupon he promptly withdrew his objections. The matter was compromised by changing the railroad time only, and during the winters of 1920-21 and 1921-22 when it was ten o'clock by "town time" in the Rhineland, it was nine o'clock railroad time.

Policy Relative to Industrial Disputes.[1]

In November 1920 Ordinance No. 53, replacing Ordinances Nos. 5 and 14, was adopted, thus modifying the policy relative to industrial disputes. With a view to the prevention of strikes boards of conciliation were to be constituted. These boards proved to be almost uniformly successful. They were appointed only in cases of disputes involving industries wherein interruptions of service would affect the forces of occupation. The efforts of one board prevented a potentially serious strike in the brown coal industry in the British zone; another settled a dispute, involving 20,000 workmen, in the coal mines near Aix-la-Chapelle; a street car strike in Coblenz was averted through the efforts of a third board. The acceptance of the findings of these boards was not compulsory, but their intervention produced good effects as both sides had faith in the impartiality of the High Commission.

Issue of Numerous Ordinances.

The ordinances and instructions published by the High Commission on January 10, 1920, were considered to constitute the minimum amount of legislation necessary to ensure the safety, maintenance and requirements of the forces of occupation. It was realized that unforeseen developments would make necessary the modification of the original ordinances and the enacting of additional legislation. From the beginning, however, the American and British representatives wished to limit the number of

[1] A.M.G., Volume I, (Chapter 10) and III, (Appendix No. 24, Page 893.)

ordinances issued by the High Commission. They deemed it wiser to solve problems, as they arose, by the application of existing ordinances, rather than to enact a mass of legislation with which the German population could not be expected to become familiar. The Belgian and French High Commissioners, influenced by their national systems, apparently desired to provide an ordinance for every situation that might arise; and as time passed their views prevailed in increasing extent. In 1920, the number of ordinances and instructions increased, respectively, from 6 to 67 and from 7 to 12. Much of the new legislation consisted of modifications of the original ordinances, but on the other hand a number of entirely new ordinances were published. The most important of the new ordinances were -

Ordinance No. 15. - Organization of working of telegraphs, telephones and wireless telegraphy.

Ordinance No. 17 - Powers of the Interallied Rhineland Navigation Commission.

Ordinance No. 18 - Interpreting Article 9 of the Rhineland Agreement and providing for the exemption from customs duties of the members and personnel of the High Commission and of the armies of occupation.

Ordinances Nos. 20 and 32 - Permitting the Administration of State Property (Reichsvermoegensverwaltung) to operate in the Occupied Territoriés and defining the competence of that organization.

Ordinance No. 29 - Right of the High Commission to veto appointments of German officials in the Occupied Territories.

Ordinance No. 49 - Billeting and quartering of personnel of the High Commission and the armies of occupation.

Ordinance No. 53 - Settlement of industrial disputes.

Ordinance No. 59 - Interpreting Article 6 of the Rhineland Agreement regarding the right of requisition.

Ordinance No. 65 - Associations and educational establishments.

Ordinance No. 67 - Sport and the transport of arms, ammunition and war material.

Article 5 of the Rhineland Agreement gave the High Commission authority to adapt the civil administration of the occupied territories to the needs and circumstances of military occupation. The High Commission accordingly modified five German laws and suspended five others. in so far as their application to the occupied territories was concerned. In addition, Ordinance No. 48 was enacted, providing that in case of conflict between German laws or regulations and legislation of the High Commission, the latter should govern.

Obstructive Tactics of German Authorities.

During the latter part of 1920 there developed among the German authorities a disposition to oppose the authorities of occupation, to contest their demands and delay compliance therewith, and to obstruct the execution of their measures. These obstructive tactics were believed to have been inspired by the Berlin Government; at any rate the obstructive attitude became especially evident immediately after certain speeches had been made in the Rhineland by members of the German Cabinet. The Chief offender was the Reichsvermoegensverwaltung (National Property Administration), a body constituted by the German Government to meet the requirements of the forces of occupation with regard to funds, housing and military establishments in general.[1] In April, 1920, upon the assurance that the purpose of this organization was to cooperate with the authorities of

[1]Volume I, Chapter 10. The Reichsvermoegensverwaltung controlled all national property in the occupied territories.

occupation and to facilitate the supply of the armies' needs, the High Commission had authorized it to begin operations.

Early in 1921, the National Property Administration demonstrated its obstructive policy by refusing to make repairs to certain military establishments and to construct others, orders for which had been given by the military authorities. Incidents of its refusal to meet the legitimate demands of the military authorities were reported from all four zones. As Article 8 of the Rhineland Agreement provided that the German Government should place all necessary facilities at the disposal of the occupying forces, the High Commission adopted Ordinance No. 69; this ordinance authorized requisitions to be made directly on Burgermeisters, or others, whenever the National Property Administration failed to supply the needs of the armies. The advisability of excluding the organization from the occupied territories was also taken under consideration.

The American policy required prompt obedience to proper orders, but also involved just and even kindly treatment of the Germans. In general the American policy had been appreciated, especially by the local German officials, who had displayed good will and a spirit of co-operation. Now, however, both the people and the officials in the American area began to use obstructive tactics, and it appeared as if they were endeavoring to ascertain the extent to which they could safely go in this regard.

The policy of obstruction culminated in a series of systematic refusals to obey orders received from the military authorities of occupation. This resulted in the suspension from office of seven officials and the preferring of charges against them.

The officials were directed to submit their defenses to these charges. An examination of the defenses showed that the charges against two of the officials could not be sustained and that the action of the other five had been the result of instructions from the President of the National Property Administration. The latter was then called upon to explain why he had acted as alleged by the five suspended officials. After a further investigation of the matter the High Commission dismissed the President and three high officials from office and expelled them from the occupied territories, the cause being their wilful refusal to comply with lawful requisition orders of the military authorities. As the other implicated officials had been acting under instructions they were not dismissed from office, but the advisability of their transfer to other administrative duties was conveyed to the German Commissioner.

After this drastic action of the High Commission, no further difficulty with the National Property Administration was experienced and the period of obstructive tactics in general came to an end.[1]

The Sanctions.

On March 8, 1921, the High Commission was informed by the Supreme Council that as the German Government had not accepted the decisions of the Allied Governments with regard to reparations, certain sanctions would be applied.[2] Among these sanctions

[1] See Appendix 4 to this Chapter.
[2] See Volume I, Chapter 1.

were the following:

> The duties collected by the German Customs Service on the external frontiers of the occupied territories was to be paid to the Reparations Commission.
> The duties would continue to be levied in accordance with the German tariff.
> A line of customs posts on the Rhine and on the perimeters of the bridgeheads was to be temporarily established. The tariffs, as to both imports and exports, to be applied on this line were to be determined by the High Commission in conformity with the instructions of the Allied Governments.

The High Commission was instructed to study the application of these sanctions, making an immediate report thereof to the Allied Governments.

Ordinance No. 77 was then adopted for the purpose of preventing the removal to unoccupied Germany of any funds derived from customs duties collected in the occupied territories. It established special regulations for the operations of the (German) Department of Customs, Department of Exports and Imports and the Office for the Prevention of the Flight of Capital. This ordinance and others dealing with the sanctions were issued by the High Commission in compliance with the decisions of the Allied Conference at London in March 1921, and not under the Rhineland Agreement. The ordinances and important decisions adopted in connection with the sanctions will be found in Appendix 5 to this chapter.

The management of the new customs service was entrusted by the High Commission to the Customs Managing Board, organized for the purpose and consisting of three members - the Belgians, British and French each having one representative. This body, together with the Economic and Financial Committees of the High

Commission, was directed to prepare a plan of the new customs system for submission to the Allied Governments. Officials of the latter immediately inaugurated a supervision of the German customs services on the external frontiers of the occupied territories.

On April 11 information was received from the Council of Ambassadors that the proposals of the High Commission had been approved. Ordinances Nos. 81 and 82, regulating the customs organization of the occupied territories and including special regulations with regard to imports and exports, were then published, and later were followed by Ordinances Nos. 84, 86, 87, 88, 89, and 91, all dealing with the new customs regime.

As soon as the new customs line was established, the German Government protested that it was a violation of the Treaty of Versailles and that the industries of the occupied territories would be ruined. Special protest was made against the requirement of licenses for shipments of goods into and out of the Rhineland. This system of licenses was open to serious objections from the Germans on three main counts -

> It caused great delay to shipments and therefore reduced the amounts of imports and exports.
>
> Licenses for the importation of luxuries or other unnecessary articles from Allied countries could be issued freely.
>
> In order to encourage trade with the Allied countries, and especially with France, licenses for imports from Germany and neutral countries could be issued at discretion.

The fears of the Germans proved to have been justified, and they charged that the issue of licenses was influenced by favoritism, bribery and other forms of corruption. When the German complaints were taken under consideration by the High Commission, the several High Commissioners admitted that discriminations in favor of the trade of Allied countries were being made. No apologies were offered, and the practice was justified on the grounds that the customs regime was one of sanctions, and the more effective the new regime proved the greater would be the injury to Germany.[1]

The economic sanctions were officially removed on September 30, but the system of licenses, in a modified form, was continued in force.

The American authorities did not participate in the application of either the economic or the military sanctions. Orders from Washington prohibited any participation, and the policy of the American Department was one of observation only. However, the American Representative held the view that the sanctions had been adopted with a view to forcing Germany to agree with the demands for reparations (the London Ultimatum) and to create a revenue to be applied to the payment of reparations; therefore the efforts of the High Commission to work harm to the trade of the occupied territories were not justified.

Occupation of Duesseldorf, Duisburg and Ruhrort.

Another sanction applied to Germany as the result of her failure to accept the London Ultimatum was the occupation of the cities of Duesseldorf, Duisburg and Ruhrort. These cities are the Rhine parts of the great Ruhr-Westphalia coal and industrial district. The occupation was completed on March 8, 1921, by a

[1]See Appendix 6 to this Chapter for the licenses granted under the system used.

large force of allied troops. French troops composed the bulk of this force and the Belgians were next in strength. A small British contingent participated in the occupation, withdrawing to Cologne within a short time thereafter.

This step gave control of all coal shipped from the Ruhr District by means of the Rhine. The High Commission was not concerned with the movement of the troops. but in the execution of its duties to provide for the safety of the armies of occupation, it published Ordinance No. 76. This Ordinance authorized the military authorities, in their discretion, to impose a censorship over the telephone and telegraph system and also to censor the press with regard to the publication of news of the movement of troops.

Subsequently the French High Commissioner proposed that the High Commission institute negotiations with the military authorities relative to an extension of its jurisdiction to the newly occupied territory which became known as the Duesseldorf-Duisburg-Ruhrort Bridgehead. The proposal was withdrawn and the military government of the new bridgehead was contined in force throughout 1921, although some of the legislation of the High Commission was applied there in a modified form.

Resignation of Herr von Starck, German Commissioner.

In the autumn of 1920 the French military authorities several times requested the German authorities to hand over to their jurisdiction a number of Germans charged with offenses against the forces of occupation. These requests were made in pursuance of Article 4 of the Rhineland Agreement which provided for the arrest and handing over to the military authorities, upon their demand addressed to the German authorities in either unoccupied or occupied territory, of any person charged with an offense against the persons or property of the forces of occupation. As none of these requests had been complied with, on February 9, 1921, the High Commission informed the German Commissioner of the matter and granted him a period of ten days within which to furnish an explanation of the delay. His reply was considered unacceptable. A marked increase in the number of attacks against French soldiers had developed, the offenders taking refuge in unoccupied Germany where they were assured of immunity. In view of these facts the High Commission formally requested the German Commissioner to inform his Government that in failing to comply with the demands of the Allied military authorities made under Article 4 of the Rhineland Agreement and, above all, in systematically subjecting such demands to delay for which there was no justification, it was in danger of forcing the High Commission to take measures to obtain satisfaction. In order to prevent the escape of accused persons into unoccupied Germany, a control over the movements of persons between occupied and unoccupied territory might be instituted; and the High Commission might be compelled to ask their governments to withdraw their consent to the appointment of the German Commissioner.

On April 13 the High Commission warned the German Commissioner that if the German authorities failed to hand over the accused persons by April 18, the High Commissioners would ask their respective governments to withdraw their consent to his appointment or even entirely to suppress the office of German Commissioner. The period of 5 days passed, and none of the accused had been surrendered. It appeard to the High Commission that the German Commissioner had been extremely dilatory in his efforts to effect the delivery of the accused persons, that the replies to the notes of the High Commission consisted merely of arguments attempting to show that in particular cases the German Government was not obliged under the Rhineland agreement to honor the requests, and finally that the attitude of the German Commissioner rendered him a hindrance rather than a help. The High Commission then addressed a note to the British, Belgian and French Governments, recommending that in accordance with Paragraph 5 of the Versailles note of July 29, 1919, they recall their approval of the appointment of Herr Von Starck as German Commissioner.

During the discussions leading up to the dispatch of this note, the French High Commissioner urged the making of a recommendation that the office of German Commissioner be abolished. He argued that the work of the High Commission could be done more efficiently through direct relations with the governments of the states having territory comprised in the occupied portion of Germany. As the other High Commissioners objected, the French High Commissioner did not press the point.

Anticipating the action of the Allied Governments, Herr von Starck resigned, leaving Dr. von Brandt, the Deputy Commissioner, to act in his stead until a successor should be appointed.

Appointment of Prince von Hatzfeld-Wildenburg.

Without delay, the German Government proposed that Prince von Hatzfeld-Wildenburg should replace Herr von Starck. On July 27 the Council of Ambassadors informed the German Government that the Allied Governments were disposed to approve its proposal, provided the following conditions were complied with:

> The new German Commissioner should endeavor loyally to cooperate with the High Commission and should refrain from inciting systematic obstructions to its measures. He should exert his influence with the German officials with a view to putting an end to the inimical attitude and lack of courtesy toward the authorities of occupation.
>
> Certain secret associations, whose object was to excite the population, would have to be dissolved.
>
> The German Government should undertake to surrender the offenders demanded by the High Commission - in spite of repeated requests these offenders had not yet been delivered to the occupying authorities.
>
> The German Government should promise to abstain from taking secret steps to nullify the judicial measures adopted in execution of the Rhineland Agreement by the Allied authorities.

On August 16 the German Government replied that it would comply with those conditions and expressed the hope that consent to the appointment of Prince von Hatzfeld-Wildenburg would be given. The Council of Ambassadors forwarded this information to the High Commission and recommended that if the German Government was ready to comply with the conditions cited, the High Commission should raise no objection to the proposed appointment.

Prince von Hatzfeld-Wildenburg arrived in Coblenz in September 1921. Shortly before his arrival, the French High Commissioner brought to the attention of the High Commission the fact that the German Government had again refused to deliver an offender against the ordinances. He proposed a recommendation to the Allied Governments that they should consider the withdrawal of their agreement to the appointment of the successor to Herr von Starck. The other High Commissioners pointed out that the German contention in the case in question - that it did not come within the purview of the Rhineland Agreement - was susceptible of argument on both sides. It was finally decided to make no objection to the assumption of office by the new appointee.

The whole incident - the resignation of the German Commissioner and the appointment of his successor - shows an interesting phase of the French policy in the Rhineland. The German Commissioner was inherently an obstacle to the furtherance of the separatist movement and to the commercial penetration of the French; his activities, doubtless, were largely directed toward the keeping alive of a National spirit amongst the population of the occupied territories and to opposing the spread of French influence. It was probably in realization of the foregoing that from the beginning the French High Commissioner sought to bring about the abolition of the office; that he did not succeed was largely due to the stand taken by General Allen and the British High Commissioner.

Removal of the Economic Sanctions.

On August 13, 1921, the Supreme Council decided to abolish the economic sanctions which had been instituted on March 7. The removal of the sanctions was to be contingent upon the acceptance of certain conditions by the German Government, one of which read as follows:

> The constitution of an interallied organization which will collaborate with the competent German authorities for the examination and delivery of licenses for the importation and exportation of goods to and from firms in the Occupied Territory of Germany, as defined by the Treaty of Versailles, the sole object of this organization being to insure that the operation of the German system does not result in setting up discriminations contrary to the provisions of Articles 264-267 of the Treaty, as far as the Occupied Territory is concerned.

The customs experts of the High Commission immediately began to draft plans for such an organization. Conflicting views developed, and the German Government displayed a disinclination to agree to the proposals of the Allies; the customs experts could not reach an agreement, and it was finally decided to submit the matter to the Supreme Council for decision.

On September 28, the Supreme Council informed the High Commission that the economic sanctions would be removed on September 30, and requested that all necessary measures should be taken. The High Commission then adopted Ordinance No. 98, repealing its customs sanctions and establishing transitional measures. One of the primary purposes of this ordinance was to insure the validity of actions taken during the maintenance of the customs regime. The Interallied organization mentioned above had not been established; hence it was also provided that until that body was ready to function, the existing organization for the issue of import and export licenses should be maintained in so far as goods crossing the external frontiers of the occupied territories were concerned.

The new interallied organization had not come into being at the end of 1921, and the old organization continued to function. Otherwise, with the publication of Ordinance No. 98 on

September 30, the Allied control of customs in occupied Germany had come to an end.

Separatism and the Smeets Incident.

Although supported by the French the Separatist movement did not develop any real strength during 1920 and 1921. There were two main obstacles in the path of the movement: the refusal of the Americans to permit a change in the Government of the American area except by constitutional means, and the strong opposition of all but a very small minority of the inhabitants of the occupied territories. Since the failures of the revolutionary attempts in 1919, both the movement and its leading exponent, Dr. Dorten, had fallen into disrepute among the Germans.[1]

The Separatist movement was kept alive, however, largely by the "Rheinische Republik", a newspaper published in Cologne. This paper was edited by Herr Smeets, a German of questionable reputation, who had once been put in jail in the Belgian zone, but had since entered into more or less friendly relations with the French authorities.

Smeets organized a meeting in the interest of the Separatist movement, to take place in Bonn (French Area) on December 4.[2] The meeting was attended by several hundred persons, and addresses were made by Smeets and others, including several Frenchmen and one Belgian. The next day Smeets was arrested by the German authorities

[1]A.M.G., Volume I, Chapter 15.

[2]The American authorities were invited to attend the meeting, but declined to be represented there in any form. It was understood that the British had pursued the same policy.

for having failed to appear in court to answer a charge, preferred some time before, of insulting a policeman. At his request, Smeets was taken before the Cologne representative of the High Commission, who examined the papers in the case. The representative stated that the papers appeared to be entirely regular and that he could not interfere; he then directed that pending a decision by the High Commission, Smeets should not be removed from Cologne.

On the same day, the High Commission met in extraordinary session. The French Deputy High Commissioner stated that the French High Commissioner, then in Paris, had instructed him over the telephone to request the High Commission to secure the immediate release of Smeets and to see that all proceedings against the latter were quashed. The French Deputy High Commission said further that French public opinion was greatly aroused over the arrest of Smeets, and he insisted that the High Commissioner comply with the French request. For the time being no action was taken by the High Commission.

The French High Commissioner hurried back to Coblenz and called another meeting of the High Commission. Upon opening the meeting, he, as President, reminded the others present that the American Representative had no vote in matters coming before the High Commission.[1] After much discussion the High Commission ordered that pending its definite decision, Smeets be released.

[1]The American Representative invited attention to the fact that except by the permission of the American Commanding General, the High Commission would exercise no authority in the American area. Such reminders as these two were very rarely made.

This action was taken in order to preserve harmony and was based upon the argument that Smeets had been arrested because of political acts begun during the Armistice and continued, after the coming into force of the Treaty of Peace.[2] The main reason for this action, however, was the understanding that the French Government had insisted upon the release of Smeets and had intimated that it would bring pressure to bear in order to effect it.

Industrial Disputes.

The boards of conciliation appointed by the High Commission settled a number of industrial disputes during 1921. Among these were strikes on the street railways of Coblenz and Munchen-Gladbach (Belgian area) and a strike of engine room crews on the Rhine.

The strike of the street railway employees in Coblenz caused some difficulty, as they struck without having complied with Ordinance No. 53. The leaders were then arrested, and the other employees were informed that they would be prosecuted unless they resumed work at once. At the same time they were informed that they were at liberty to strike after having complied with Ordinance No. 53. The striking workmen immediately resumed work, and shortly afterward the board of conciliation effected a compromise.

In the strike of the engine room crews the High Commission was confronted with a more difficult problem. The crews had declined to accept the wage scale and working hours proposed by the shipping concerns and the German authorities.

[2]See Ordinance No. 70.

Portions of the right bank of the Rhine were in the neutral zone and many of the strikers resided in unoccupied territory; hence it was impracticable to proceed against them for failure to comply with Ordinance No. 53. Nevertheless, the High Commission informed the striking crews that the ordinance must be observed and that it would not tolerate a waterways strike or a wilful stoppage of traffic. The strikers were told that if they failed to heed this warning, they might be brought to trial. The strike was finally settled when both sides accepted conditions which had been formulated in large part by a board of conciliation.

There were two serious railroad strikes in Germany during 1921. The first strike, which occurred in January, did not spread to the occupied territories, as the High Commission had informed the railway personnel that a strike in the Rhineland would not be permitted, since the maintenance, safety and requirements of the forces of occupation would be endangered thereby. In December, during the second strike, a number of workmen disobeyed orders and stopped work. Their services were requisitioned by the military authorities, after which there was no further trouble on the railways in the occupied territories.

Additional Legislation.

Among the important ordinances enacted during 1921 were the following:

Ordinance No. 69 - Contributions to be made by the German Government in pursuance of Article 8 of the Rhineland Agreement.

Ordinance No. 71 - Espionage, etc.

Ordinance No. 90 - Right of the High Commission to divest the German authorities and courts of certain cases which concerned the Allied forces.

In addition three German laws were suspended in so far as they applied to the occupied territories.

APPENDICES.

A P P E N D I X 1

Internal Organization and Constitution

of the

H I G H C O M M I S S I O N

I. THE HIGH COMMISSION.

1. All decisions in matters within the competence of the High Commission shall be made in the name of the High Commission.

2. All documents officially embodying or conveying decisions made in the name of the High Commission shall whenever the High Commission shall so decide, bear the signatures of the High Commissioners. They shall otherwise be signed by the President of the High Commission and countersigned by the Inter-Allied Secretaries General. In cases of urgency, when the President is absent, one of the High Commissioners shall sign for him. Correspondence of minor importance may be signed, by order of the High Commission, by the two Inter-Allied Secretaries General.

3. No High Commissioner of any one nation nor any member of his staff shall have the right to take decisions on matters within the competence of the High Commission alone.

II. THE COMMITTEES.

1. The High Commission shall appoint such Committees as it shall consider necessary.

2. A Committee shall normally consist of one delegate of each High Commissioner. If necessary, the delegate may be assisted by technical advisers. The Inter-Allied Secretaries General shall have the right to attend the meetings of the committees.

3. The duties of the committees shall be: -

   (a) to examine matters sent them for study and report, or as an exceptional measure for decision, by the High Commission;

   (b) to examine and report direct on matters within certain categories laid down by the High Commission;

   (c) as an exceptional measure to take decision on certain matters within other categories similarly laid down by the High Commission.

4. Matters within categories (b) and (c) mentioned above shall be submitted direct to the Committees by the Inter-Allied Secretaries General. A list of such matters shall be attached to the agenda of the next Sitting of the High Commission, accompanied by a copy or brief analysis of the relevant documents. Urgent matters shall be specially marked by the Inter-Allied Secretaries General. The High Commission may give instructions to any Committee without awaiting its report.

5. As far as possible, and except in cases of emergency, a High Commissioner shall not bring forward at a sitting of the High Commission matters within the categories (b) and (c) mentioned above, but shall cause the revelant papers to be passed to the Inter-Allied Secretaries General, who will forward them on a file, together with all necessary documents, to the Secretaries of the Committee concerned, to be placed on the Agenda of the Committee.

6. Decisions of a Committee taken upon matters within its competence shall go out in the name of the High Commission.

7. Decisions of Committees shall be taken by unanimous vote. Upon the request of any one member any question brought before the Committee shall be referred to the High Commission.

8. Any one member of a committee may present a separate report to the High Commission, explaining his personal views on any question brought before the Committee.

9. As far as possible, any proposal brought before the High Commission by a Committee shall be submitted in the definite form in which the Committee considers the decision, note, despatch or telegram should be sent out should its proposal be accepted.

III. THE INTER-ALLIED SECRETARIES GENERAL.

1. The High Commission shall appoint two Inter-Allied Secretaries General, one French speaking and one English speaking.

2. (a) One French speaking and one English speaking member of the Inter-Allied Secretariat General shall be appointed to act as Inter-Allied Secretaries General whenever it is impossible for the Inter-Allied Secretaries General to perform their duties.

(b) Whenever it shall be necessary he shall take the place of the Inter-Allied Secretary-General of his language and shall sign for him.

3. Subject to the President and the High Commissioners the two Inter-Allied Secretaries General shall be responsible for:-

(a) the internal working of the joint Inter-Allied services of the High Commission

(b) the co-ordination of the work of the Committees;

(c) the disposal of all matters presented for the consideration of the High Commission, the reception and distribution of all correspondence or documents intended for the High Commission, and the preparation and despatch of all correspondence or documents emanating from the High Commission;

(d) the preparation of the Agenda and the Minutes of the High Commission;

(e) the keeping of the Archives of the High Commission.

A. (I) The Inter-Allied Secretaries General shall supervise the working of the Inter-Allied Secretariat General and the Secretariat of the Committees.

(II) The Inter-Allied Secretaries General shall be authorized to submit for the approval of the High Commission or to institute themselves such internal measures of working as they shall consider will conduce to the prompt and efficient despatch of the business of the High Commission.

B. (I) The Inter-Allied Secretaries General shall be held responsible for seeing that the decisions of a Committee, insofar as they are not referred to the High Commission, do not conflict with previous decisions of the High Commission or of the Committees.

(II) In doubtful cases they shall withhold their signatures and refer the point in question back to the Committee concerned, or bring it to the attention of the High Commission or of the President.

C. (I) The Inter-Allied Secretaries General may annotate any document submitted to the High Commission or to a Committee with a view to mentioning any previous rulings on, or references to, the question therein raised.

(II) They shall present a weekly statement to the President of the matters which are in suspense before the High Commission or before the Committees.

Outgoing correspondence.

(III) The heading of documents signed by the four High Commissioners, or of the correspondence of minor importance signed by the two Inter-Allied Secretaries General, shall be "The Inter-Allied Rhineland High Commission to ... ."

(IV) The heading of documents signed by the President shall be, in conformity with diplomatic usage, "The President of the Inter-Allied Rhineland High Commission to....... ..." The text shall be drawn up in a way to show that the President is expressing himself "in the name of the High Commission."

(V) Documents provided for in paragraph (IV) immediately above shall be signed on the right by "The President of the Inter-Allied Rhineland High Commission" and countersigned lower down on the left by "The Inter-Allied Secretaries General". In the case of the President's absence, and where there is urgency, the document shall be signed for the President by one of the High Commissioners.

(VI) Correspondence of minor importance shall be signed by the Inter-Allied Secretaries General alone under the formula "By Order of the High Commission, the Inter-Allied Secretaries General."

(VII) The Ordinances of the High Commission shall be communicated to the Army Commanders by, or by order of, the Allied Commander in Chief.

Communications between the High Commissioners and the various military authorities shall always take place through the channel of the Allied High Command.

(VIII) The communications of the High Commission shall be drawn up in French. They shall be drawn up in English when they are addressed to an American or British Authority or individual. Communications addressed to the Inter-Allied High Command shall be accompanied by an English translation.

(IX) Communications from the High Commission to any Inter-Allied Commission or Committee shall be addressed according to the usage in force.

(X) Copies of all important letters despatched in the name of the High Commission shall be sent to the four High Commissioners for their information.

(XI) Except as may in the future be otherwise prescribed, all communications from a Committee of the High Commission or from any authority whatsoever of the High Commission shall go out under the formulae and following the procedure laid down above.

<u>Incoming Correspondence.</u>

(XII) All communications for the High Commission shall pass immediately to the Inter-Allied Secretariat General.

(XIII) Communications for a particular High Commissioner or member of his staff should be expressly addressed to him.

(XIV) Communications intended in fact for the High Commission, but addressed or delivered to a particular High Commissioner, to a member of his staff or to a committee of the High Commission, shall be handed over immediately to the Inter-Allied Secretariat General.

(XV) Communications which only contain information shall be circulated or distributed to the High Commissioners by the Inter-Allied Secretaries General.

(XVI) Communications which require action shall be treated by the Inter-Allied Secretaries General as follows: -

(a) In principle the said communications shall be placed on the Agenda of the High Commission.
(b) Communications which fall within the categories (b) and (c) of paragraph 3 of Part II above, and which should be sent for the study, or as an exceptional measure, for the decision of the Committees, shall be forwarded without delay to the Secretaries of the Committee concerned.
(c) Reply can be given direct by the Inter-Allied Secretaries General to the questions of detail on matters of which the principle has already been settled by an earlier decision of the High Commission.

D. Agenda and Minutes of the High Commission.

(I) The Agenda for a sitting of the High Commission shall be decided by the President and distributed at least twenty-four hours before the sitting takes place. In consequence, documents to be placed on the agenda shall be received in the Inter-Allied Secretariat General at least two days before the sitting of the High Commission at which they are to be discussed.

(II) The Minutes shall be drawn up in French and in English. They shall in principle consist of:-

(a) a brief statement made in impersonal form of the material facts presented for the consideration of the High Commission, and a reference to the relevant documents.
(b) the decision of the High Commission. When a High Commissioner desires any statement made by him to appear in the Minutes he should expressly say so at the sitting.

E. The Archives of the High Commission.

(I) All incoming documents received by the High Commission shall be placed in the Inter-Allied Archives.

(II) All documents (or copies of them) which the High Commission or any of its Committees takes into consideration when reaching a decision, shall be placed in the Inter-Allied Archives.

(III) Copies of all documents sent out in the name of the High Commission shall be placed in the Inter-Allied Archives.

IV. <u>THE SECRETARIES OF COMMITTEES.</u>

1. Each Committee shall ordinarily possess one French speaking and one English speaking secretary who shall be members of the Inter-Allied Secretariat General.

2. All questions forwarded for the consideration of the Committee shall be sent, together with all necessary documents, by the Inter-Allied Secretaries General to the Secretaries of the Committees, to be placed on the Agenda of the Committee.

3. In principle the Secretaries of the Committee shall not receive direct communication of any documents forwarded for the consideration of the Committee, and shall transmit any documents wrongly so received in the Archives, to be sent to the Inter-Allied Secretaries General, nor, except in case of urgency, shall a member raise a question in Committee of which notice has not been given on the Agenda.

4. The Secretaries of the Committee shall be responsible for the preparation, adoption and proper distribution of the Minutes of the Committee.

5. The Secretaries of the Committee shall be responsible for the preparation of the documents sent by the Committee to the Inter-Allied Secretaries General for reference to the High Commission.

6. The Secretaries of the Committee shall be responsible for the preparation of any letters embodying decisions of the Committee, which, accompanied by the file containing all relevant documents and a minute signed by both Secretaries, giving the decision of the Committee, shall be sent by them to the Inter-Allied Secretaries General for signature and despatch.

7. The Secretaries of the Committee shall be responsible for the deposit in the Inter-Allied Archives of all documents taken into consideration by the Committee.

8. The Secretaries of the Committee shall present

a weekly statement of the matters which are in suspense before the Committee to the Inter-Allied Secretaries General for transmission to the President.

V. PROCEDURE FOR THE SITTINGS OF THE HIGH COMMISSION.

1. The High Commission shall sit on the day and at the time fixed by it at the end of each Sitting.

2. The President, or in his absence one of the High Commissioners, can call an extraordinary sitting of the High Commission.

3. Notice shall be sent to the High Commissioners of an extraordinary Sitting by the Inter-Allied Secretariat General.

4. The President shall be the High Commissioner of the French Republic. In his absence, the other High Commissioners shall preside in turn.

5. The High Commissioners, or in the case of absence of one of them, his Deputy, shall alone have the right to vote at the Sittings of the High Commission.

6. Technical Advisers and heads of departments may be brought to attend by each High Commissioner, and the High Commission may summon and give hearing to any person it considers necessary.

7. The Inter-Allied Secretaries General, or in their absence the acting Inter-Allied Secretaries General, shall act as Secretaries at the Sittings of the High Commission. They shall be assisted by such a number of clerks or typewriters as they judge necessary.

8. The President shall declare the Sitting open. The proceedings shall begin by the adoption of the Minutes of the last Sitting, with or without modification.

9. The President shall then bring forward the various paragraphs of the Agenda in such order as, in agreement with the High Commission, he shall consider advisable.

10. The President shall give the right of speech to the members of the High Commission, in the order in which it has been asked. He may give the right of speech to the persons referred to in paragraph 6 above.

11. (a) Decisions of the High Commission shall be taken by a majority of votes.

(b) Each High Commissioner shall have one vote, but in the case of equality of votes, and when the High Commissioner of the French Republic is present, the President shall have the right to give a casting vote.

12. With regard to the above paragraph, the dissenting High Commissioner, or High Commissioners, may appeal to their Governments, but such an appeal shall not, in cases of urgency delay the putting into execution of the decisions taken, which shall then be carried out under the responsibility of the members voting for the decisions.

13. When a decision shall have been taken it shall be recorded in the Minutes.

14. When the matters on the Agenda shall have been finished, the High Commissioners or their Deputies may raise a question and ask either that it shall be put on the Agenda of the next Sitting, or in urgent cases, be immediately discussed.

15. The President shall declare the Sitting raised.

M E M O R A N D U M

regulating

THE ATTRIBUTES OF THE ALLIED MILITARY AUTHORITIES AND OF THE INTER-ALLIED RHINELAND HIGH COMMISSION.

(Text approved by the Council of the Principle Allied Powers, June 13, 1919.)

I.

Each High Commissioner shall be directly responsible to his own Government. Economic questions shall be brought up in the first instance by the High Commission before the economic Supreme Council, as long as it exists.

II.

The ordinances of the High Commission shall be

communicated to the Commanders of the armies by the Allied Commander-in-Chief, or by his order.

III.

When the High Commission enacts ordinances which affect the interests of the forces of occupation, and which were not inspired by the military authorities, it must, before so doing, consult the aforementioned authorities.

IV.

Relations between the High Commission and the different military authorities shall always be effected through the Allied High Command.

V.

The civil or official commissions which exist at present or are to be nominated by one or more of the Allied or Associated Governments and which are concerned with the civil administration or economic life of the civilian population in the Occupied Territory shall be placed, if maintained, under the authority of the High Commission.

VI.

a) The nomination of each of the High Commissioners shall be submitted to the other Allied and Associated Governments represented.

b) The French member of the High Commission shall be president.

c) Decisions shall be taken by a majority vote.

d) Each High Commissioner shall have one vote, but in the case of an equality of votes, the President shall have the right to give a casting vote.

e) In either of these cases, the dissenting High Commissioners may appeal to their Governments, but, in urgent cases, such an appeal shall not delay the application of the decisions taken which shall then be carried out on the responsibility of the members having voted in the affirmative.

VII.

When promulgating decrees or proclamations or when intervening, in any other way, in the civil administration during the state of siege, the Commander-in-Chief shall continue to act together with the High Commission and only with the approval thereof. It is understood that this procedure shall not apply to measures of a purely military character.

(Initialed) W.W.
G.C.
D.L.G.
S.S.

APPENDIX 2.

Ordinances, Instructions and Important Decisions of the HIGH COMMISSION in 1920 and 1921 – as Contained in the Official Bulletins

Note: The Official Bulletins are in a separate volume which accompanies those containing the text.

APPENDIX 3.

Kreis Representatives.

Duties of Kreis Representatives.

Protests of German Authorities against the Powers of the Kreis Representatives.

Definition of the Functions of the Kreis Representatives.

Policy of the Various Departments of the High Commission with Regard to Kreis Representatives.

Organization of Kreis Representatives in the Respective Zones.

Assumption of Duty by Kreis Representatives in the American Zone.

KREIS REPRESENTATIVES

In the course of the negotiations in the month of July, 1919 at Versailles, regarding the interpretation of the Rhineland Agreement, it was agreed that with the ratification of the Treaty, there would no longer be any administrative or supervising officials attached to the German authorities as was the case during the Armistice. It was further agreed that the High Commission could maintain fixed representatives with the duty of securing liaison between the local German administration, the local military authorities and the High Commission itself.

It was in view of this agreement that the High Commission, in its ordinances and instructions, provided for certain duties to be performed by its representatives. These duties are as follows:

1. Investigation of appeals for clemency. (Ordinance No. 2)

2. Transmission of reports from German courts concerning cases with respect to offences against the ordinances of the High Commission or against the persons or property of the armed forces of the Allies. (Ordinances Nos. 2 and 57)

3. Investigation of complaints of civilians against the armed forces of the Allies. (Ordinance No. 2)

4. Supervision of political meetings. (Ordinance No. 3)

5. Issuing of arms permits to German officials. (Ordinance No. 3 and Instruction No. 7)

6. Report on publications of a nature to prejudice public order or endanger the security or the dignity of the High Commission or of the troops of occupation. (Ordinances Nos. 3, 13, and 97)

7. Surveillance of the circulation of German military persons, German nationals from unoccupied territory and from points outside of Germany and all foreigners. (Ordinance No. 3)

8. Supervision of stocks of arms and ammunition in the hands of dealers. (Ordinance No. 3)

9. Reports on persons becoming permanent residents of the Occupied Territories. (Ordinance No. 3)

10. Reports on appointments and changes of German officials. (Ordinances Nos. 29 and 54)

11. Supervision of the display of flags. (Ordinance No. 30)

12. Reports on the food supply. (Ordinance No. 39)

13. Transmission of reports of payments made by the German authorities for billets placed at the disposal of the personnel of the High Commission and the members of the Allied armies. (Ordinance No. 49)

14. Transmission of requests for the establishment of Boards of Conciliation to settle industrial disputes. (Ordinance No. 53)

15. Supervision of associations and educational establishments in order to prevent military training. (Ordinance No. 65)

16. The receipt and transmission with observations of appeals from arrest or seizure by reason of political acts or commercial transactions during the period of the Armistice. (Ordinance No. 70)

17. Countersigning of passports of Allied nationals. (Ordinance No. 73)

18. Report on carrier pigeons. (Ordinance No. 79)

19. Report on the spread of venereal disease. (Ordinance No. 83)

20. Receipt and transmission with observations of appeals in customs cases arising out of the customs sanctions. (Ordinance No. 98)

21. Transmission to the military authorities of applications by the German authorities for the use of the Allied military forces in order to re-establish order. (Instructions 1 and 9)

22. Receipt of reports from the German authorities concerning public order, disturbances, strikes, industrial unrest, public meetings, elections and other matters which might affect public order. (Instruction No. 2)

23. Supervision of the execution by the German authorities of German sanitary police regulations. (Instruction No. 4)

24. Liaison between the German authorities and military authorities.

25. Inspection of prisons. (Instruction No. 14)

26. Compilation of monthly coal report for the information of the Coal Committee. (Minute 240 of 13th sitting of the High Commission)

27. Inspection of obsolete arms in possession of Germans and issuance of permits to retain same. (Minute 259 of the 15th sitting of the High Commission)

28. Control of import and export of ammunition, fire arms, explosives and war material; approval of sporting arms permits issued by the German authorities. (Ordinance No. 67)

29. Reports on local food and fuel situation. (Instruction No. 2 and Ordinance No. 39)

The general duty of the Kreis representatives is to see that all ordinances, instructions, orders and regulations promulgated by the High Commission and the military authorities are observed and enforced.

Very soon after the ordinances and instructions of the High Commission were published in January, 1920, the German authorities objected that the High Commission had conferred on its Kreis representatives a series of administrative powers of extraordinary and decisive importance. The German contention was that it was not within the role of a liaison officer to supervise and forbid meetings, to suspend newspapers and to supervise the carrying of arms, inasmuch as these powers were all of an executive and administrative nature. The Germans also protested that the High Commission had violated the promise given by the Allies to the effect that no supervising officials would be attached to the German authorities. The German Government always contended that the representatives of the High Commission should confine themselves simply to expediting transactions between departments and to transmitting the requests of one department to another. They should have no right to interfere in the internal affairs of the German authorities, to check their records, to supervise their orders and to issue instructions to them.

The German authorities later proceeded to file a series of protests against alleged violations of authority by the representatives of the High Commission. The most important of these protests are the following:

It was alleged that the representative of the High Commission in the district of Ludwigshafen required that all public meetings besides political meetings, should be notified 48 hours in advance.

The representatives of the High Commission at Kreuznach and Langenschwalbach were alleged to have demanded the right to participate in meetings of German self-governing bodies, such as the Kreisausschuss and the Kreistag.

The representatives at Trier, Euskirchen and Aachen were alleged to have required a detailed report from the German authorities on a great number of purely economic questions with which the authorities of occupation were not concerned.

The representative at München-Gladbach was alleged to have required reports from the chamber of commerce, although German chambers of commerce are representative bodies of an economic character and cannot be considered as government, state or municipal authorities.

The representative at Kreuznach was alleged to have requested the German authorities to issue police regulations concerning the suppression of infectious diseases.

The representative at Trier was alleged to have stated that he would be obliged to order the dismissal of certain German police and customs agents unless they performed their duties in a more satisfactory manner.

In reply to these protests, the High Commission stated that they were unfounded and were formulated only in order to bring to discussion the powers of its representatives. However, it saw fit to define the powers of its representatives as follows:

The representatives of the High Commission are liaison officers between the German authorities and the population on the one hand and the High Commission and the military authorities of occupation on the other. They are the usual intermediaries in all relations between the High Commission and the German authorities and the population.

The role of liaison officer qualifies the representatives to transmit to the German authorities and to the population all ordinances, instructions and decisions of the High Commission and on their own initiative to address to the same persons all requests for information within the meaning of the ordinances, instructions and decisions of the High Commission.

The representatives of the High Commission are competent to transmit to the German authorities the orders, instructions and requests for information of the military authorities.

The representatives of the High Commission had no powers of administration or control which would enable them to interfere in German administration by approving or countersigning official documents.

It will be observed that the protests set forth above are directed almost exclusively against French representatives. Although the German authorities protested against the general extension of the powers of the representatives of the High Commission in all the zones, there were only a few cases of protests against the representatives in the Belgian zone and none at all with respect to the conduct of the representatives of the British and American zones. This fact may be attributed in part to the natural animosity that exists between the French and German people, but is probably due in large part also to the tendency of the French representatives to interfere in local administration. This tendency is not disconnected from the separatist movement and the desire to create a buffer state in the Rhineland.

The policy of the American and British Departments has always been to limit the functions of the Kreis representatives to those of liaison officers, as had been agreed with the German Peace Delegation. The policy of the French Department and to some extent also of the Belgian Department has been to try to extend the jurisdiction of the representatives. This tendency has been restrained but it will be noted that almost every new ordinance carries with it an additional duty and therefore an extension of power for the representatives of the High Commission.

A striking instance of interference with German administration is furnished by the action of General de Metz, chief representative of the High Commission in the Palatinate. In the course of Communist disturbances at Speyer in September, 1921, General de Metz relieved the Regierungspraesident, and the chief of the German police and made known publicly that the town police were incompetent. He did this by publishing a proclamation and a decree. The proclamation stated that in view of incidents that had taken place, he felt compelled to assume responsibility for the maintenance of public order and that the German police force would be placed under the command of the French authorities of occupation. The decree specified that the German police authorities of the district of Speyer should, until further orders, be under the command of the Town Major of Speyer; that the municipal police of Speyer should be dismissed until further orders, since their incapacity to perform any duties whatsoever had been duly proved; and that public order would continue to be maintained by the French authorities of occupation.

The action of General de Metz was justified neither by the Rhineland Agreement nor any of the ordinances and instructions of the High Commission. In accordance with Instruction No. 1, he should have forwarded the request of the Chief of Police for the use of the Allied military forces in order to re-establish order, to the military authorities in charge of the Allied troops concerned. In accordance with Article 13 of the Rhineland Agreement, the military authorities would then have had the power to take such temporary measures as might have been necessary for restoring order. Under no circumstances did General de Metz have the right to issue the proclamation or the decree referred to above. The German authorities immediately protested against the action of General de Metz.

The High Commission thereupon informed General de Metz that it was not his duty but that of the military authorities to take measures to restore order in accordance with Article 13 of the Rhineland Agreement. In the meantime, General de Metz had withdrawn his orders and permitted the German authorities to resume control of the police forces.

Although this is an exceptional case, it may well be used to illustrate the French policy of showing the population that the authorities of occupation and not the German authorities are in control of the Occupied Territories.

In general, the High Commission has a representative in every Kreis, but this rule has been departed from, notably in the American zone. The American zone comprises 8 Kreise, which lie wholly within the Occupied Territories, and 3 Kreise which lie partly in the Occupied Territory and partly in the neutral zone. Representatives are maintained in the following Kreise:

Stadtkreis Coblenz.
Cochem
Montabaur
Mayen
Ahrweiler
Neuwied

These representatives function directly under the supervision of the district representative at Coblenz. There is only one district representative in the American zone because that zone consists of only one district.

The Belgian zone consists of 18 Kreise. In principle, there is one representative for each Kreis. The zone is divided into two districts, known as the Regierungsbezirk Aachen, comprising 6 Kreise, and the Regierungsbezirk Düsseldorf, comprising 12 Kreise. Each district is supervised by a district representative, who in turn reports to a superior representative.

The British zone comprises 8 Kreise. In principle, each Kreis has one representative, who functions directly under the district representative. The zone consists of one district, supervised by a district representative, known as the Cologne Commissioner.

The French zone of occupation comprises 51 Kreise, which lie wholly within the occupied area, and 2 Kreise which lie partly in Occupied Territory and partly in the Saar Valley. In principle, each Kreis has one representative, who is directly under the district representative. The zone includes 2 provinces, 2 districts and two groups of Bezirke called districts by the French, in each of which there is a delegate superior, who reports directly to the delegate general.

During the last few months of the Armistice period, the Inter-allied Rhineland Commission maintained unofficial representatives in the various Kreise in the Occupied Territories. These representatives performed no official duties and limited their activities to observation of the functions of the officers in Charge of Civil Affairs. Their task was to familiarize themselves with the duties they would be called upon to perform at the time when the High Commission came into power. When the High Commission assumed authority, on January 10, 1920, its representatives began to function officially in the Belgian, British and French zones. In the American zone, it was not quite clear what duties the Kreis representatives should perform as the Commanding General of the American Forces in Germany considered it advisable to maintain an officer in Charge of Civil Affairs in each Kreis for the time being.

On January 17, 1920, Major General Allen reached an agreement with Mr. P. B. Noyes wherein the duties of the Kreis representatives and the officers in Charge of Civil Affairs were defined. It was agreed that the representatives of the army and of the High Commission should jointly inspect German prisons. This procedure was adopted in order to insure that the German authorities concerned might take no advantage of the peculiar situation to raise any questions as to the respective authority of the army and the High Commission. German officials should give one copy of all reports to the Officer in Charge of Civil Affairs and one copy to the the representative of the High Commission. Notification of meetings should be given to the military authorities and in case of political meetings, the Kreis representatives should be at liberty to attend with the representative of Civil Affairs. Permits to carry fire arms should be issued by the Provost Marshal instead of by the Kreis Representative. Applications by the German authorities for the use of the Allied military forces in order to reestablish order should be sent to the Officer in Charge of Civil Affairs instead of to the Kreis Representatives.

Instructions to this effect were sent to the Officers in Charge of Civil Affairs on January 20, 1920. The Officers in Charge of Civil Affairs were at the same time informed that it was particularly desirable that German officials be given no opportunity to raise any question as to the respective authority of the army or of the representative of the High Commission in any Kreis, and that the Officers in Charge of Civil Affairs should confer with the representatives of the High Commission freely and frankly on any questions that might arise. All officials were to be informed that any instructions received from the representatives of the High Commission falling within the ordinances of the said commission, must be observed the same as military orders emanating from the Commanding General (1).

By General Order No. 19, dated March 4, 1920, Instruction No. 7 of the High Commission was published for the information and guidance of all persons whom it concerned. The publication of this instruction meant that the control of permits to carry fire arms passed from the hands of the military authorities to the representatives of the High Commission. On June 7, 1920, the authority of the

1. Vol. 4, pp. 416-420

Kreis representatives was further extended when they were permitted to supervise economic and political questions as defined by the ordinances.

On October 21, 1920 the Officers in Charge of Civil Affairs in the Kreise were relieved, their duties being taken over by the representatives of the High Commission. Since that date, the Kreis Representatives in the American zone have had the same authority and performed the same duties as the representatives in the other zones.

APPENDIX 4.

Action of the High Commission with Respect to German Officials.

Power of the High Commission to Remove and Veto the Appointment of German Officials.

Protest of the German Authorities against This Power.

Number of Officials Removed from Office in 1920 and 1921.

Number of Vetoes of Appointments of Officials in 1920 and 1921.

Attitude of the Respective Departments of the High Commission with Regard to Officials.

ACTION OF THE HIGH COMMISSION

WITH RESPECT TO

GERMAN OFFICIALS

Article 5 of the Rhineland Agreement provides that the German authorities shall be obliged, under penalty of removal, to conform to the ordinances issued by the High Commission for the purpose of securing the maintenance, safety and requirements of the Allied and Associated forces. In the course of the negotiations at Versailles in June and July, 1919, the German Peace Delegation protested against this provision, stating that it was unnecessary because all German officials would obey ordinances lawfully issued. It was further requested that if it were found necessary to dismiss an official, the dismissal should take place through the German Commissioner, who would be responsible for the examination of the case.(1) The Allied reply was that the request put forward in the German note would involve an alteration in the text of the Agreement. It was further admitted that, except in cases of urgency, officials might on the orders of the High Commission be dismissed without undue delay either by the German Commissioner or by the competent German authorities. The High Commission reserved in all cases its right itself to dismiss officials whenever necessary. (2)

In accordance with the provisions of Article 5 of the Rhineland Agreement referred to above, the High Commission issued an ordinance, number 29, regarding the right to veto the appointments and remove German officials in the Occupied Territories. This ordinance provides that the High Commission may veto the appointment

(1) Volume IV, Appendix 51, page 126.

(2) Volume IV, Appendix 51, page 139.

of any German official designated to serve in the Occupied Territories if, in the opinion of the High Commission, such action is necessary for securing the maintenance, safety and requirements of the Allied and Associated forces. This ordinance further provides that any German official serving in the Occupied Territories may be removed from office by order of the High Commission if, in the opinion of the High Commission, such action is necessary for securing the maintenance, safety and requirements of the Allied and Associated forces, or when such official fails or refuses to conform to the ordinances of the High Commission.

The German authorities immediately protested against this ordinance, claiming that the appointment of officials was a right which belonged only to the German authorities and that the right of veto assumed by the High Commission constituted interference in the administrative sovereignty of the German Government in the Occupied Territories. Exception was taken especially to the provision that officials might be removed if they failed to conform to the ordinances of the High Commission. In this respect the German authorities urged that German officials were only bound to submit to the instructions of German authorities superior to them and that it was not within the province of the High Commission to prosecute officials who conformed with instructions received from their superiors.

The High Commission replied that the German Peace Delegation had admitted that the right of dismissal was vested in the High Commission and that it followed that the High Commission had the power to veto the nomination of officials whose introduction might stir up disorder. The German request for the withdrawal of Ordinance 29 was therefore not complied with.

In accordance with Ordinance 29, twenty-three officials were removed from office during 1920 and 1921. Sixteen of these were removed at the request of the French authorities, four at the request of the American authorities, two at the request of the Belgian authorities and one at the request of the British. The four officials whose removal was requested by the American Department were Herren Klett, Strasser, Müller and Maurer. They were all officials of the Administration of State Property (Reichsvermogensverwaltung) and had refused to obey lawful orders received from the military authorities. The circumstances attending their removal have been described above. (1)

During 1920 and 1921 the appointments of twenty -three German officials were vetoed by the High Commission; nineteen of them at the request of the French Department and nine at the request of the Belgian Department. The American and British Departments did not request the veto of any appointments during this period.

From these figures it will appear that the French and Belgian authorities exercise a much closer supervision over German affairs than is the case with the American and British authorities. There is a tendency on the part of the French representatives in particular to supervise even unimportant administrative acts and to report trifling errors of omission or commission with the recommendation that the official concerned be admonished, punished or removed. Such an attitude naturally generates friction and is possibly responsible for the unwillingness of German officials to cooperate with the authorities of occupation. The attitude of the American, and British authorities toward German officials has been to treat them strictly and even severely, but at all times fairly and

(1) See above, page 27.

impartially. The result is that there have been far fewer cases calling for disciplinary action against German officials in their zones.

During the latter part of 1921 the authorities of occupation noticed an ever increasing tendency of the German authorities to appoint officials who were not natives or residents of the Occupied Territories. The reason adduced by the German authorities for this increase was that a large number of Catholic officials had lost their positions through the cession by Germany of certain territories whose population was almost exclusively Catholic, notably Alsace-Lorraine, Posen and parts of Silesia. The Rhineland being preponderantly Catholic, it was considered advisable to transfer Protestant officials from the Rhineland and replace them with Catholics. The authorities of occupation, however, were of the opinion that this influx of non-resident officials was to be ascribed to a desire to maintain Prussian influence by means of appointing reactionary Prussian officials. With this in mind, the authorities of occupation began to veto a large number of officials, especially school teachers who came from East Prussia and Silesia. In the case of school teachers this action may be justified in that it prevents the spread of reactionary ideas among children. It is to be anticipated that the number of appointments that are vetoed in 1922 will be greatly in excess of the number for 1921.

A P P E N D I X 5.

Ordinances and Important Decisions with Regard to Customs Regime is Published in the Bulletin of the Customs Managing Board.

Note: The Bulletin mentioned as published in a separate volume which accompanies those containing the text.

A P P E N D I X V

Table of Licenses Granted from April 20, 1921 - to - December 30, 1921.

Licenses granted for imports to and exports from the Occupied Territory during the sanctions, from April 20th, 1921, to September 30th, 1921.

(Amounts in paper Marks).

	Imports	Exports.	Trade Balance. Debit.	Credit.
France...........	1,057,997,948	394,373,292		663,124,656
England..........	653,821,878	672,862,482	19,040,604	
Belgium..........	818,536,423	480,774,058		337,762,365
Italy............	77,858,986	142,359,917	64,500,931	
United States....	554,717,884	222,512,079		332,195,805
Holland........	757,699,585	1,768,286,430	1,010,586,845	
Switzerland......	52,768,812	211,767,460	158,998,648	
Spain............	74,330,948	106,147,928	31,816,980	
Other countries..	327,246,115	1,579,412,032	1,252,165,917	

APPENDIX 7

Action of the High Commission with Respect to the German Press.

Authority of the High Commission to Prohibit, Exclude and Suspend Publications.

Number of Publications Prohibited, Excluded or Suspended during 1920 and 1921.

---oOo---

ACTION OF THE HIGH COMMISSION

WITH RESPECT TO THE GERMAN PRESS.

In Ordinance No. 3, the High Commission provided that all newspapers, pamphlets or publications of a nature to prejudice public order, endanger the security or the dignity of the High Commission or of the troops of occupation, were forbidden, and if published, might be seized by order of the High Commission, or in case of emergency, by order of the representative of the High Commission in the Kreis. In the case of a daily publication, the representative of the High Commission in the Kreis might order its suspension or exclusion from the Occupied Territories for three days. The High Commission might also order newspapers or periodicals to be suspended or excluded from the Occupied Territories for a period not exceeding three months.

In enacting this ordinance, it was not the intention of the High Commission to subject German publications to censorship prior to their publication. It was, however, considered wise to make provision for any cases of slanderous attacks affecting the armies or the High Commission.

In 1920, it was found necessary to exclude, suspend or prohibit 42 publications; in 1921, such action was taken with respect to 55 publications. In 1920, such proceedings were taken in 34 cases at the request of the French; in 4 cases at the request of the English; in 3, at the request of the Belgians; and in one, at the request of the Americans. In 1921, action was taken in 45 cases at the request of the French; in 6 cases, at the request of the English; and in 4 cases at the request of the Belgians. During this year, the American authorities did not find it necessary to ask for prohibition, suspension or exclusion of any publication.

The above figures indicate that most of the requests for action on the part of the High Commission emanated from the French authorities. This resulted from the fact that the German press has conducted a systematic campaign against the French troops of occupation Almost all the evils of the occupation are attributed to the French, while the other armies are left in comparative peace, probably as a result of orders from Berlin. The colored troops who form a large part of the French army of occupation are the favorite subject for attacks in the press, and these attacks have been responsible for the majority of suspensions, prohibitions and exclusions of publications.

It can hardly be said that the maintenance, safety or requirements of the armies of occupation are seriously endangered by the statements that appear in German publications, but the High Commission has adopted the attitude that slanderous statements might cause friction between the population and the troops and thereby cause a disturbance of public order. If this is not taken into consideration, it will necessarily appear that the High Commission, has, at times, been unduly severe in its attitude towards the press. In September, 1921, the High Commission found it necessary to publish Ordinance No. 97, modifying its ordinance concerning the press. The principle new provision is to the effect that any periodical publication which has been suspended or excluded more than once may, in the event of a subsequent offence, be suspended or excluded for an indefinite period. This action was necessary because a number of publications which had been excluded or suspended several times, persisted in their attacks; it, therefore, appeared that suspension or exclusion for three months was insufficient.

APPENDIX 8

INDUSTRY AND LABOR.

INDUSTRY AND LABOR[1]

Industrial labor troubles were expected to increase at the beginning of 1920 owing to the rapid depreciation in the value of the mark and the consequent increase in the cost of living and the almost prohibitive price of clothing. Added to this was the fact that in many important industries wage agreements were due to terminate on January 1 and January 31, 1920. New demands were made by employees for higher wages and bonuses with which to meet the increased cost of living. In some cases the men went on strike before the expiration of the existing wage agreements and before bringing their cases to the Boards of Arbitration as normally prescribed by German law. In other cases strikes were started by workers against the wishes of their own leaders who no longer had control over their men.

The most important strikes in the Occupied Territories in the early part of 1920 took place in the British zone. One of these strikes was that of the laborers employed in the railway workshops at Cologne. This strike began on January 2d and spread to the other railway workshops in the British zone. It arose out of the demand of the workmen for 3.30 M. per hour instead of 3 M. which they were receiving. It was found necessary for both the British authorities and the German railway authorities to issue proclamations calling upon the men to return to work because, under the terms of the Armistice, railway employees were subject to military law, as the safety of the Armies depended upon the operation of the railways. This strike was settled when it became known that the German Railway Administration was willing to

[1]A.M.G., Volume I, Chapter 10.

grant the increase demanded.

A more important strike was that of the workers in the brown coal mines. The cessation of work by the miners meant that a number of plants supplying the Belgian and British zones of occupation would have to cease to operate. The authorities of Occupation were, therefore, directly affected. This strike was engineered by agitators and was largely political. Attempts were made to compel men who were willing to work to go on strike. This led to serious conflicts and the military authorities were called upon to intervene. A number of arrests were made by the British authorities who also issued a proclamation calling upon the men to resume work in view of the suffering they were inflicting on thousands of innocent people who were depending upon them for a supply of coal. The leaders of the Miner's Union agreed that the strike was unjustified and succeeded in persuading the strikers to resume work.

Much unrest was created among railway employees throughout the Occupied Territories when the German Railway Administration announced its intention to close down railway workshops whose output was not satisfactory until the workmen agreed to accept the piece-work system of pay. It was found necessary to take this action because production had shown a continuous downward curve since the revolution of 1918 and no efforts had succeeded in raising it. A number of workshops were closed as a result of this decision but the laborers almost without exception accepted the new conditions and returned to work.

In the early part of 1920 there was much unemployment in the Occupied Territories but the situation there was not nearly as serious as in Unoccupied Germany; moreover the number of unemployed in the Occupied Territories constantly decreased with the lapse of time.

The chief causes for unemployment were lack of coal and raw materials. In the latter part of 1919 the return of former prisoners of war slightly increased the number of unemployed but this factor was never very important as the men soon found employment. The shortage of fuel, however, had very serious consequences. A number of large factories had to close down their works temporarily owing to coal shortage and many firms could only work short hours or three to four days a week for the same reason. The shortage of raw materials was felt particularly in the steel and textile industries, although these industries were also affected by the lack of rolling stock and transport facilities.

During February 1920, the labor situation improved very considerably. This improvement may be ascribed to a number of causes. In many of the more important industries new wage agreements were concluded at the end of January and almost invariably increased wages and bonuses were granted. Moreover, the workmen began to show a greater willingness to work. This was specially the case with the coal miners who on their own accord undertook to work extra shifts in view of the national necessity for increased output of coal.

The continued rises in prices of foodstuffs and clothing carried with them considerable increases in wages but, generally speaking, wages did not increase in proportion to the cost of living. The average increase in the cost of living at the beginning of 1920 over the cost of living in 1914 was about 800%; the increase in wages, however, amounted to only about 550%. From this it is apparent that the workmen were in most cases justified in asking for further increases.

Although the cost of production had increased enormously as compared with pre-war days, the employers of labor in large industries were usually able to meet the demands for high wages without fear of bankruptcy, as in many cases they had accumulated vast profits during the war and were still making large profits in those industries which were producing goods for export.

The Kapp revolution and the following disturbances had very little influence on industrial life in the Occupied Territories except insofar as the supplies of coal and essential products were interrupted by the almost complete cessation of railway traffic in the adjoining Ruhr district There was very little sympathy among the laboring classes for the Communists who were opposing the Reichswehr. There were also a number of demonstrations in various parts of the Occupied Territories against the short-lived new government established by Kapp. A large number of laborers in the districts adjoining the Ruhr Basin ceased work, crossed the frontier and helped the Communists but they returned to work as soon as they saw that the movement was a failure.

A serious situation for German industry arose in the summer of 1920 through the very substantial improvement in the value of the mark. In the early part of the year the mark had fallen as low as 100 to the dollar. The result was that the prices of raw materials. foodstuffs, clothing and practically all other articles rose tremendously. It was only natural that the workmen should demand high wages and receive them. When the mark improved in June to 35 and 40 to the dollar, there was no pronounced decrease in prices. This placed German manufacturers in an unfavorable position in their relations with foreign countries. Many concerns were unable to compete with foreign firms because of high wages and the high price of manufactured goods. The cancellation of foreign orders crippled the smaller factories who had no reserve capital or stocks. The larger factories were able to continue operating with a reduced output because they had stocks of raw material which they worked off and stored in the hope of an improvement. The closing of many factories and the reduction in the number of working hours brought about a substantial increase in unemployment. Other factors that affected industry disadvantageously at this time were the shortage of water power caused by an unusual drought and the lack of coal in sufficient quantities to supply the deficiency. The trades which suffered most were the metal, furniture, and leather industries. The big factories engaged in making bridges, engines, machinery, tubes, and other heavy goods were not so seriously affected because they had contracts for long periods ahead and also large reserves of coal.

The situation was still further complicated by the enactment on June 25 of a law providing for the deduction of 10% of a man's wages as income tax to the State. The enforcement of this law called forth many protests from the workmen and in some localities brought about a series of strikes. In several cases important manufacturing concerns were forced by the workmen under threats of strike to pay the income tax themselves. There were also instances where the workmen immediately demanded a 10% increase in wages. The attending unrest together with the conditions discussed above resulted in a partial stagnation of industry for several months. At a later date the income tax was slightly modified so that it was less objectionable to the workman. This change and a general increase in wages helped to allay the discontent.

In August 1920 the employees of an important electrical power plant in the British zone of occupation demanded an increase in pay and, as the increase was not granted immediately, went on strike. This plant supplied power to factories scattered over a very large area. The industrial life of the surrounding district and the safety of the Armies of Occupation were dependent upon it. The I. A. R. H. C. informed the representative of the employees that a strike in this plant could not be permitted until certain regulations had been observed, because the operation of the plant was of vital necessity to the safety of the Armies and the maintenance of public order. The men were ordered to return to work and to bring up their case for settlement before a Board of Conciliation in the prescribed manner. Instead of heeding this admonition the men decided to continue the strike. A number of the ring leaders and agitators were thereupon arrested by the authorities of Occupation. Shortly afterwards the workmen accepted the compromise offered by the employers and the strike came to an end. This is one of the few instances in which German workmen deliberately disobeyed orders given by the authorities of Occupation.

The improvement in the value of the mark in the early summer proved to be of short duration and the autumn months showed another drop. The result was a resumption of activity in the industries (manufacturing for export, but the industries) dependent upon raw materials purchased abroad and producing for inland consumption were placed at a disadvantage. There was immediately a rapid increase in the cost of living with the result that the demands for increased wages again became numerous. In most cases the mens' demands were partly or wholly granted with the net result of an increase of about 10% over the existing scale for the first half of the year. The smoothness with which most of these wage disputes were settled indicates that the working classes were becoming tired of the perpetual agitation of the extremist element and were desirous of seeing settled conditions.

The prospects for both industry and labor appeared to be good at the beginning of 1921. In the majority of enterprises wage increases had been granted and even if the laborers were not as well off as before the war, they had escaped from the misery to which they were reduced in the summer of 1920. While all the industrial countries of Europe were struggling in a serious economic crisis, industrial and commercial conditions in Germany were improving. This situation was attributed to the fact that Germany could manufacture and sell at prices far below those prevailing in other countries. Wages although rather high in marks remained far below wages paid in countries with a more favorable rate of exchange. Furthermore the price of coal in Germany was only about sixty percent of the price in adjoining countries and it was becoming easier to procure raw materials. In fact, the year began with a decrease in the price of almost all raw materials. Manufacturers especially those producing for export, anticipated a period of prosperity.

However, this hope was not destined to be realized at once. As early as February there was a decrease in production with corresponding increase in unemployment. Strikes again became more numerous and the reduction of working hours assumed serious proportions. It was obvious that the universal economic crisis had reached Germany. Orders placed by foreign firms in the previous year had been filled and were not renewed. As time went on the situation became worse because the employers could not or would not comply with the demands of the workmen.

It was at this inauspicious moment that the customs sanctions established by the London Conference came into effect. It was expected that the erection of a customs frontier between Occupied and unoccupied Germany would have a most disastrous influence on industry. It can not be denied that industry was adversely affected but there was no economic disaster as had been prophesied by the German press. Until the end of June the situation remained unfavorable. In the textile and metal industries which are among the most important in the Occupied Territories, working hours were reduced considerably. The tobacco industry was crippled not only because of the new customs line but also because of the imposition of new taxes.

The progressive and continuous decline of the mark which began in earnest in July gave a new impulse to industry. July saw a renewal of activity in plants manufacturing railway material, metal products, shoes, and textiles. This resumption of work helped to alleviate the problem of unemployment which had assumed serious proportions. Employers again began to call upon the workmen to work overtime, and skilled labor was at a premium. This period of comparative prosperity lasted throughout the rest of the year.

In the month of November economic activity reached its highest level as a result of the fall of the mark. December marked the first stage of a period of depression which had been prophesied for several months. Enterprises which were dependent upon foreign trade noticed that orders from abroad were becoming scarce. The rush to buy in Germany which had been the feature of the market in October and November began to slacken. Furthermore, a lack of coal and coke, especially in the metal industry, together with an increase in the price of raw materials, compelled some establishments to reduce their production. The end of the year was therefore characterized by a slight reduction of economic activity and by increasing uneasiness concerning the immediate future. There were, however, no indications of an industrial crisis and there was no reason to believe that the slight industrial depression would become more serious.

In the course of 1921 strikes were particularly numerous. In the Belgian zone alone, there were forty strikes involving more than five hundred laborers each, making a total of sixty thousand strikers. In addition to actual strikes, there were a large number of threatened strikes which were obviated by Boards of Conciliation appointed by the German authorities and by the Interallied Rhineland High Commission.

One of the most serious strikes occurred at the coal mines in the district of Moers. This strike was brought about largely by agitators from unoccupied Germany who sought to cause

A P P E N D I X 9

AMERICAN POLICIES

AMERICAN POLICIES

by

MANTON DAVIS

American Legal Adviser

Interallied Rhineland High Commission

---o---

The outstanding feature of American action on the Interallied Rhineland High Commission was its disinterestedness. This was recognized by Germans and Allies alike. America had no territorial, political, economic or financial purposes to serve through its representation on the High Commission. This representation from the beginning was unofficial. The representative was not a High Commissioner but an "unofficial observer." Major General Henry T. Allen, commanding the American Forces in Germany, was from June, 1920, the American Representative.

The influence General Allen exercised on the High Commission resulted not only from his country's disinterestedness and his own but also from the fact that he had under his jurisdiction a definite part of the occupied territories.

When the Treaty of Peace came into effect and the Rhineland Commission began to function, it became General Allen's delicate duty to determine how conflicting authority in the American zone of occupation might be harmonized. The other nations participating in the occupation were at peace with Germany while American continued to be technically at war. Those nations forthwith began to apply and to be governed by the Treaty of Peace and the Rhineland Agreement while America continued in occupation under the Armistice Convention. The Rhineland Agreement contemplated no such confused state of affairs. It did not provide for the contingency that one of the occupying powers might not accept the Treaty. As against the Germans, the Rhineland Commission, under the Rhineland Agreement, had jurisdiction over the whole of the occupied territories including the American zone. As against the Germans, the American Commanding General, under the Armistice Convention, had jurisdiction to govern the American zone. No instrument defined the powers of these authorities as between themselves.

General Allen announced in the beginning that insofar as it was practicable the American area would be governed by the same laws and ordinances and in the same manner as the other areas were governed. This decision made necessary a larger participation in the deliberations of the High Commission than the position of unofficial observer apparently justified. The High Commission accepted the situation with the best of spirit and always accorded to the views of the American Representative the same weight as if he had been an official member.

The guiding principle of American policy was to permit the Germans to govern themselves without interference except insofar as military occupation made interference necessary. This principle was declared in the Rhineland Agreement and appears therein as Article 5. The Rhineland Agreement and the government of the occupied territories by the civilian commission it established had been reluctantly accepted at the Peace Conference. As the occupation continued, not infrequent efforts were made in the High Commission and by the military authorities to extend control beyond the limits set by the Rhineland Agreement. American effort was consistently directed toward restraining these encroachments, at the same time rigidly exacting from the German authorities and people obedience to established laws and ordinances. Disobedience or defiance of constituted authority on the one side and harsh measures applied in excess of treaty-given right on the other would inevitably have created and perpetuated ill will between the nations concerned. General Allen's influence was always exerted toward preventing so unhappy a result. Specific incidents illustrating the application of this influence follow:

I. THE RHINELAND REPUBLIC - THE CASE OF DR. DORTEN.

In 1919 there was concocted a revolution contemplating the establishment of a separate Rhineland republic. Its capital was to be Coblenz. Proclamations to this effect were issued and distributed. Separation of the Rhineland from Germany had been urged at the Peace Conference as a measure necessary to the safety of the Allied nations. The advance party of this embryo government came to Coblenz. The American Commanding General caused the leaders of this party to be arrested and they were tried and convicted for attempting to incite public disorder. This ended that effort to establish a Rhineland republic. The leading German spirit in this movement was a certain Dr. Dorten of Wiesbaden. In the proclamations issued he was named as the President of the Rhineland Republic.

The incident above related amongst others gave rise to certain observations at the Peace Conference. In a reply to the German note regarding the occupation of the left bank of the Rhine it was stated:

> "On the other hand, Allied and Associated Governments, who are anxious to avoid the introduction of a disturbing element in the occupied territories, can not admit that legal proceedings should be instituted by the competent German authorities in respect of political or commercial acts relating to the armistice period, when these acts have not given rise to legal proceedings on the part of Allied and Associated authorities."

- (Documents relative to the control of the Rhineland territories during the military occupation, paragraph 9. Referred to hereafter as "Versailles Notes").

One of the first ordinances promulgated by the High Commission provided that without the consent of the High Commission no judicial proceedings should be instituted against any person in the occupied territories for any administrative or political act done during the period of the armistice. (Ordinance 2, Article 31.) This ordinance prevented the Germans from prosecuting Dr. Dorten for treason.

On 24th July 1920 armed police from Frankfurt in the unoccupied territory abducted Dr. Dorten and took him out of the occupied territories. This act, of course, was a violation of the authority vested in the High Commission as the supreme representative of the Allied and Associated powers in the occupied territories. The High Commission took strong and immediate action. The Regierungspraesident of Wiesbaden and the Deputy Chief of Police of Wiesbaden were dismissed from their respective offices and expelled from the occupied territories for their supposed guilty knowledge and probable complicity in this transaction.

General Allen sent for the Commissioner representing the German government with the High Commission (Reichskommissar) and told him that this action was an attempt to flout the authority of the High Commission and would probably result in serious consequences if the German government did not immediately cause Dr. Dorten to be returned to his home. General Allen at the same time advised the Reichskommissar that the German government would be well advised to apologize to the High Commission. The Reichskommissar received this advice with poor grace at first, but after a long discussion accepted it and acted upon it. As a result of this advice Dr. Dorten was returned to his home and on 6th August 1920 Herr von Starck, the Reichskommissar, appeared formally before the High Commission and stated that by direction of the German government he expressed to the Rhineland Commission the regret of the Central German government that, contrary to the ordinances in force in occupied territory, the apprehension of Dr Dorten had taken place.

General Allen thereupon advised the High Commission that in his opinion the German government had made proper amends and that since it did not appear that the Regierungspraesident of Wiesbaden was directly concerned in this transaction, the High Commission should permit him to return to the occupied territories and should reinstate him in his office. After much delay and discussion the High Commission at its 60th meeting, on 1st November 1920, ordered that the Regierungspraesident be reinstated.

II. THE RAILWAY ORDINANCE.

The Rhineland Agreement, Article 10, provides that the personnel employed on all railways shall obey any orders given by or on behalf of the Commander-in-Chief of the Allied and Associated armies for military purposes. Among the first ordinances promulgated by the High Commission was Ordinance No. 6 concerning railroads. This ordinace was drafted prior to the coming into force of the Treaty of Peace. Its preparation involved much discussion and many meetings. It was argued that the railroads of the occupied territories should be controlled by the military authorities: that the only way to ensure that military orders to railroads would be obeyed would be to put the railroads entirely under military control. This position was stated at the meeting of 31st December 1919 as follows:

> "I think it is not a good position to say that the Germans can come to the High Commission and dispute an order given by the military authorities. I think they should first be compelled to obey it. I think the High Commission should say to the armies: 'It is your business to do so, go ahead and run the railroads'."

The American contention was that the only proper authority which the armies could exercise over the railroads resulted from Article 10 of the Rhineland Agreement and must necessarily be "for military purposes". Attention was called to paragraph 3 of the Fourth Versailles Note, which reads:

> "It is understood that the civil administration of railroads shall be exercised by the German authorities, with the reservations provided for by Article 10 of the Agreement of June 28th."

The High Commission finally accepted this point of view and consequently the railroad ordinance was drafted on the theory that the armies of occupation might not control the railroads except to issue orders "for military purposes".

III. LOCAL REPRESENTATIVES.

The High Commission maintained a local representative in each Kreis (practically a county). These officials were contemplated by the First Versailles Note paragraph 29, in which it was stated that they would have no duty of supervising the German authorities in the local administration, but that the High Commission might have the power of maintaining such representatives to secure liaison between the German administrations, the local military authorities and the High Commission itself. When the scheme for the organization of the occupied territories was first discussed one of the High Commissioners announced his intention of having military persons perform this service who would report through the military hierarchy so that reports would reach the High Commission through the Army Chief of Staff.

The American contention was that the High Commission was a civilian body and that its local representatives should be civilians under no military control. The British High Commissioner stated that his local representatives in some instances would be military officers but that all of them would be under the sole control of the High

Commission and in no sense under the control of the military authorities.

After much discussion it was determined that the Kreis representatives whatever their status, whether military or civilian, should be under the control of the High Commission in the performance of their duties and that if they were military persons they would be subject to military authority only for military discipline.

The Kreis representatives were the eyes and ears of the High Commission in the outlying districts. The High Commission was often embarrassed by the fact that in certain areas these representatives were military persons and more disposed to promote the policies of the military authorities than those of the High Commission. Had the contention prevailed that those representatives should report to the High Commission only through military channels and ultimately through the Army Chief of Staff, much delay would have resulted and many matters important to the High Commission would never have come to its attention at all.

IV. MILITARY ORDERS.

During the Armistice period each of the armies issued orders for the control of civilian officials and the civilian population. In some zones this control was exercised in minute detail. In the First Versailles Note, paragraph 24, it was stated that

> "It is the intention of the Allied and Associated authorities to regard the various decrees issued by the military authorities in the occupied territories during the armistice as having lapsed after coming into force of the Treaty of Peace. Nevertheless, it belongs exclusively to the High Commission to decide on the necessary transition measures."

The High Commission was organized really, though informally, long prior to coming into force of the Treaty of Peace on January 10, 1920. On the day the treaty went into effect the High Commission promulgated a number of ordinances and instructions. Despite this argument was made in the High Commission that while all military orders inconsistent with the ordinances of the High Commission should be deemed to have lapsed, yet all of those military orders which were not inconsistent with those ordinances should be considered still in effect.

The High Commission at its third meeting, January 14, 1920, decided that all military orders should be deemed to have lapsed and invited the military commanders to make suggestions at the earliest possible moment to the High Commission respecting matters concerning which new ordinances should be issued. In spite of this ruling, again and again it was called to the attention of the High Commission that the military authorities in certain zones were continuing to enfore their military orders for the government of the civilian authorities and population. Not only this, but it was again and again argued that the military authorities after the coming into force of the Treaty of Peace had the power to issue orders in cases not covered by the ordinances of the High Commission. The American contention, always to the contrary, prevailed and became the accepted position of the High Commission.

In making their contentions the American authorities were embarrassed no little by the situation of the American Army, which continued to occupy its area under armistice conditions after the coming into effect of the Treaty of Peace and maintained de jure its right to govern its area. General Allen originally took the position which he consistently maintained that, regardless of his legal position, he would publish the ordinances of the High Commission in his area and would govern the area in every practicable way exactly as the other areas were governed.

V. CENSORSHIP.

In the First Versailles Note, paragraph 21, it was stated:

> "Freedom of communication by letter, telegraph and telephone will be re-established between the occupied territories and unoccupied Germany, subject to a general reservation of the rights of the High Commission, etc."

After the Treaty of Peace came into force the military authorities frequently requested the High Commission to give to them the power of postal censorship over large areas of the occupied territories, asserting that this was necessary for the safety of the armies. The first application was from the French Army and was considered by the High Commission on 4th February 1920. The request was for the power to institute a general censorship in the districts of Wiesbaden and Ludwigshafen. The French representative stated:

> "General Degoutte says that his command is in danger; that he can not be responsible for the safety of his army unless permitted to take necessary measures to foresee and forestall these dangers."

The American position was that a postal censorship was an easy and a lazy way for army intelligence police to do their work but that it would be at the expense of the convenience and liberties of the people and that it would do much more harm than good; that a special censorship over the mail of a particular person or persons should be authorized by the High Commission whenever a necessity for it was shown to exist, but that no general censorship should be instituted except under circumstances where the High Commission might find it necessary under Article 13 of the Rhineland Agreement to declare a state of siege. In this case, of course, control would pass to the military authorities who could do substantially as they pleased.

This contention consistently prevailed and became the established policy of the High Commission.

VI. GERMAN LAW CONCERNING WORKMEN'S COUNCILS.

The High Commission examined all new German laws prior to their going into effect in the occupied territories. Whenever the maintenance, safety or requirements of the armies would apparently be prejudiced by any such law, the High Commission forbade or modified the law as occasion required.

On 4th March 1920 there came before the High Commission a newly enacted law concerning workmen's councils in industries (Betriebsraetegesetz). This law provided for an advisory council in every business organization. which council should have rather large rights to examine into the affairs of the company and to advise concerning the relations between employers and employees, both being represented on the council. This was one of the body of new German laws more or less socialistic in tendency. It was strongly contended that the High Commission should forbid the application of this law in the occupied territories. The power of the High Commission to forbid or control German law was, of course, limited to the needs and circumstances of military occupation and could properly be exercised only for the maintenance, safety and requirements of those armies. The contention was that this law was an experiment and its application would doubtless result in disorder to the prejudice of the safety of the armies.

The American contention was that the safety of the armies would probably be much more prejudiced by the anger aroused should the High Commission interfere in the application of a law which the German working people regarded as highly beneficial and even necessary to themselves. It is probable that the reason for opposing this law was that it was socialistic in the tendencies and that its application near to the Allied frontiers might cause trouble in the Allied countries.

The High Commission decided that it would not interfere.

VII. VETO OF THE APPOINTMENT OF OFFICIALS BY REASON OF THEIR PAN-GERMANIST TENDENCIES.

From the beginning the High Commission required that German authorities notify it of the appointment of German officials who were to serve in the occupied territories. Along with the notification the High Commission required that it be furnished with certain information concerning each appointee. It was contended in the High Commission that if a person were a Pan-Germanist this was sufficient to cause a veto of his appointment. The first conspicuous case to come before the High Commission was that of a certain Dr. Momm, appointed Regierungspraesident of Wiesbaden. It was said that this man had very pronounced Pan-Germanist ideas and that he had a son-in-law who had been a U-boat commander. During the period of the American occupation at Trier, Dr. Momm was the Regierungspraesident of the Bezirk and the Americans found his conduct satisfactory. The American Representative contended that being a Pan-Germanist was, from the German point of view, merely being a patriotic German and that such views should not be a sufficient reproach to justify a veto.

This matter came before the High Commission on 5th March 1920. Up to that time no question had ever been put to a vote in the High Commission. Every decision had been taken unanimously. The High Commission decided by a vote of three to one not to oppose the appointment of Dr. Momm. The French representative expressed his regret but stated that under orders he was compelled to record his opposition. In order not to break the solidarity in the decisions of the High Commission this case was at once reconsidered and ordered postponed. It came up for discussion again on 16th April 1920. The French representative then stated that he very much regretted that the strongly expressed recommendation of the French High Commissioner and the French Army should not be accepted by the High Commission; that he was unable to vote with the High Commission but to prevent the matter from appearing on the minutes as having been decided against the French vote he requested permission to withdraw from the agenda of the High Commission his recommendation concerning the veto of Dr. Momm. This suggestion was accepted.

For a long time this action of the High Commission was followed as a precedent but later the opposite view prevailed and Pan-Germanist views were sufficient to bring about a veto.

VIII. ARREST IN THE OCCUPIED TERRITORIES BY ALLIED MILITARY AUTHORITIES OF GERMANS CHARGED WITH THE COMMISSION OF WAR CRIMES.

After the coming into force of the Treaty of Peace the Allied military authorities arrested several Germans in the occupied territories whose names appeared on the lists of the Allied governments compiled in accordance with Article 228 of the Treaty of Peace. This article of the treaty provided that the German government should hand over to the Allied powers all persons accused of war crimes. A conspicuous example of this action occurred in the American area where the American military police at the request of the Belgian military authorities arrested a certain Dr. Hermann of Neuwied and delivered him to the Belgians who sent him into Belgium for trial. This was a mistake on the part of the American police. The American authorities at once set about trying to undo the error.

The High Commission received a protest from the Reichskommissar against the arrest of these persons. The Reichskommissar contended that the right of arrest and the duty of delivering persons accused of war crimes was in the German government and that the mere fact of the presence of the Allied armies of occupation in Germany did not change the German government's duties or responsibilities under this article of the treaty and that nothing in any treaty permitted the Allied armies of occupation arresting such persons. One of the High Commissioners proposed replying to von Starck's letter saying that the matter was not within the competence of the High Commission and that if so advised the Reichskommissar or the German government might take these questions up with the governments concerned.

The American contention was that the arrest of such persons was no business of the High Commission but that the conduct of the armies of occupation in the occupied territories was the business of the High Commission since the High Commission was named in the Rhineland Agreement as the supreme representative of the Allied and Associated governments in the occupied territories; that, since the Rhineland Agreement made the ordinances of the High Commission law alike to the German authorities and to the military authorities, the High Commission could not entirely avoid responsibility for unauthorized military action. This contention prevailed and the High Commission issued orders to the armies of occupation that no further arrests should be made by them of persons charged with war crimes. This policy thereafter continued without interruption.

Dr. Hermann above referred to was removed to Belgium and was tried and by the lower court convicted. Pending an appeal strong representations were made to the Belgian government who responded that they had no power to release a person held by the judicial authorities of Belgium. The case was subsequently reversed on appeal, remanded to the trial court and, if the writer is correctly informed, a nolle prosequi was there entered whereupon the authorities of Belgium returned Dr. Hermann to Neuwied.

IX. THE REMOVAL OF HERR VON STARCK AS REICHSKOMMISSAR AND THE OPPOSITION TO THE APPOINTMENT OF A SUCCESSOR.

When the Treaty of Peace came into effect and the Interallied Rhineland High Commission entered upon its duties there was accredited to it by the German government a civil commissioner called the "Reichskommissar". This official represented not only the Central Government, but also the several German states, the territories of which, in whole or in part, were occupied according to the provisions of Part XIV of the Treaty of Peace. This office was mentioned in paragraph 5 of the First Versailles Note, 29th July 1919. The office was not referred to in the Treaty of Peace or in the Rhineland Agreement, but in the Versailles note aforesaid the consent of the Allied and Associated governments was given to the appointment of such an official.

The Reichskommissar was of great assistance to the High Commission. His activities, however, had impeded the separatist tendencies so long and so earnestly encouraged by certain of the Allied authorities. Herr von Starck was personally not well qualified for his duties. He was of the old-school Prussian type, not very tactful and much given to legal arguments asserting that the decisions of the High Commission could not be performed.

From the very beginning, however, it was apparent that it was not so much the official as the office to which there was objection. Such a case was built up against Herr von Starck, he assisting, that no reasonable opposition could be made to the demand that the High Commission should request the Allied governments to withdraw their consent to his appointment. At the same time insistence was made that the High Commission should recommend that the office be abolished.

General Allen very strongly opposed this demand. He argued that the Reichskommissar represented not only the Central Government, but also the six states occupied in whole or in part; that inconvenience and confusion would attend any effort to transact business separately with these different states; that the district officials of the High Commission, through whom some of these states would be compelled to approach the High Commission, would be disposed to decide on their own responsibility questions of grave concern which might never come to the knowledge of the High Commission at all; that such decisions had been taken in several cases and like decisions would often be taken in the future were the several German states compelled to address the High Commission through the district officials; that helpful discussions of questions important to good relations between the Allied authorities and the German authorities and population had often been held between the Reichskommissar and the High Commission or the various High Commissioners; that these discussions would be manifestly impossible were seven German states separately represented; that the work of the High Commission would be multiplied to confusion if parallel correspondence were carried on with seven separate states; and that separate relations would injuriously affect the unity of the Central German government and would encourage the separatist movement.

This argument prevailed. Herr von Starck resigned but the governments concerned after discussion between themselves consented to the appointment of Prince Hatzfeld-Wildanburg as Herr von Starck's successor.

X. RAILWAY STRIKE IN THE OCCUPIED TERRITORIES.

In the early part of January 1922 a general strike was called on the railways of the occupied territories. Ordinance No. 53 of the Interallied Rhineland High Commission required that before any strike or lockout might occur in any public utility which in any manner served the armies of occupation, the parties should bring their controversy to the attention of the Interallied Rhineland High Commission to the end that a board of conciliation, appointed by it, might attempt to bring the parties together. This strike was called without complying with this ordinance. Before the strike order actually went into effect the High Commission was assured both by the railroad directions and by the strikers themselves that every train would be operated which the military authorities designated as being necessary for their requirements. All parties further assured the High Commission in advance that there would be no disorder.

The High Commission on 4th January 1922 convened representatives of the railroad managements and also representatives of the strikers and told them that the strike would be forbidden and would not be permitted until the provisions of Ordinance 53 were complied with. The representatives of the workers responded that the strike order had been issued and that if they then attempted to recall it, the strike would nevertheless go on and the only consequence of such an effort would be that they would lose their positions; they repeated the assurances previously made that there would be no disorder and that every train would be operated which the military authorities or the High Commission should designate.

taken in these particular cases may serve to illumine and to disclose the guiding policies of the American Representative on the Interallied Rhineland High Commission.

APPENDIX 10

Biographies of the Commissioners and Their Deputies.

American Department:	Major General Henry T. Allen Colonel David L. Stone Mr. Manton Davis Mr. P. B. Noyes Mr. W. H. Day
Belgian Department:	Baron Rolin-Jaequemyns M. Fernand Cattoir Count Raoul de Liedekerke
British Department:	Sir Harold Stuart Mr. Malcolm Arnold Robertson Lord Kilmarnock Lt. Col. Rupert Sumner Ryan
French Department:	M. Paul Tirard M. Amedee Roussellier
German Commissioners:	Herr von Starck Prince von Hatzfeld-Wildenburg

(These biographies were prepared by the respective departments and have not been changed in any respect.)

MAJOR GENERAL HENRY T. ALLEN.

General Allen was born at Sharpsburg, Kentucky, on April 13, 1859. He obtained his early education in the schools of his native state and at the Peekskill Military Academy in New York. (In 1898 he received the degree of Master of Arts from Georgetown College in Kentucky and in 1920 the degree of Doctor of Laws from the Lincoln Memorial University.) He was then appointed to the United States Military Academy at West Point, from which he graduated on June 13, 1882, as a Second Lieutenant and was assigned to the 2nd Cavalry. During the years 1885 and 1886 he was engaged in explorations in Alaska, the details of which he published in book form. On June 22, 1889, he was promoted to the grade of 1st Lieutenant, and during 1889 and 1890 he was an instructor at the United States Military Academy.

From 1890 to 1895 General Allen served as Military Attache to the American Embassy in St. Petersburg. In 1897 he was appointed Military Attache to the American Embassy in Berlin, remaining there until the outbreak of the Spanish-American War in 1898. Returning to the United States, he was appointed Major and Adjutant General; but he accompanied the American forces to Cuba and commanded a troop of the Second Cavalry in the field throughout the Santiago campaign. After his service in Cuba he served a short time as Adjutant General of the Department of the Gulf at Atlanta, then he returned to Berlin, and in November 1898 was promoted to the grade of

Captain in the Regular Army. In 1900 he was sent to the Philippine Islands as Lieutenant Colonel of the 43rd Infantry and for several months in 1901 he was Governor of the Island of Leyte. In July of that year he was assigned to the duty of organizing the Philippines Constabulary, becoming its first Chief. In 1903 he was given the grade of Brigadier General, and Chief of Constabulary by special act of Congress. In April 1907 he was promoted to the permanent rank of Major. He remained Chief of the Philippines Constabulary until 1907 when he returned to the U. S. and served in the Yellowstone Park and in Arizona. He was called to Washington in 1910 and appointed to the General Staff in charge of the Cavalry Section, where he remained for four and one half years. He was promoted Lieutenant Colonel in August 1912. In August 1914, he was sent to Europe on the U.S.S. "Tennessee" to assist in the return to the United States of the numerous Americans to whom the outbreak of the World War had caused great difficulties. After his return to the United States he served with the 11th Cavalry and on July 1, 1916 he was promoted to the grade of Colonel. With his regiment he participated in the Punitive Expedition which entered Mexico in 1916 under the command of General Pershing.

Shortly after the entry of the United States into the World War, he was promoted to Brigadier General in the Regular Army and organized a cavalry brigade at Fort Riley, Kansas. In August 1917 he was promoted to Major General in the National Army and in September was assigned to command the 90th Division at Camp Travis, Texas. He organized and trained this division, brought it to France and commanded it during its service in the Toul Sector, in the

St. Mihiel offensive and in the Muese-Argonne campaign. Immediately after the conclusion of the Armistice Agreement, General Allen was assigned to the command of the 8th Army Corps at Montigny-sur-Aube, France, where he remained until April 1919 when he was transferred to the command of the 9th Army Corps at Nogent-en-Bassigny. Less than a month later he took command of the 7th Army Corps at Wittlich, Germany In July, 1919 he was selected to command the American Forces in Germany, relieving Lieutenant General Hunter Liggett. In addition to his Military duties, in June 1920 General Allen assumed the duties of American Representative with the Inter-Allied Rhineland Commission, vice Mr. P. B. Noyes relieved. In July 1920 General Allen received his promotion to the rank of Major General in the Regular Army.

General Allen has been awarded the following campaign medals and decorations:

Spanish War.
Philippine Campaigns.
Mexican Punitive Expedition.
Distinguished Service Medal.
Grand Officer, Legion of Honor (France).
Croix de Guerre with Palm (France).
Grand Officer, Order of Leopold of Belgium.
Croix de Guerre, with Palm (Belgium).
Crank Gordon, Order of the Crown of Italy.
War Cross (Italy).
Grand Officer, Order of the Oak Crown of Luxembourg.
Order of Danilo of Montenegro (Pla ue).
La Medalla de la Solaridad, First Class (Panama).

COLONEL DAVID L. STONE.

Colonel David L. Stone, General Staff, was born at Stoneville, Mississippi, August 15, 1876. He was educated at the United States Military Academy, from which institution he graduated on April 26, 1898. He saw active service in Cuba and during the Filipino Insurrection; then assisted in organizing civil government in the Philippine Islands.

He participated in many engagements with Generals Lawton, Wheaton and Funston. In the campaign against the Moros on the Island of Mindanao, he was wounded in action. At a later date, he was in charge of the Quartermaster construction work at Fort Crook, Nebraska and at Fort Omaha. He also built Fort Sill in Oklahoma and Camp Lewis at American Lake, Washington.

During the World War, he served on the General Staff of the Third Division in France, participating in the major engagements at Chateau Thierry, Saint Mihiel and Meuse-Argonne. At a later date, he was appointed Deputy Chief of Staff of the Second Army.

As a result of his distinguished services, he received a Third Division Citation, the Distinguished Service Medal, the Croix de Guerre with palm and was made an officer of the Legion of Honor.

During the American occupation of Germany, Colonel Stone was at first on duty with the First Brigade, A.F. in G., and in January, 1920, he was assigned as Military Advisor to the American Commissioner on the Interallied Rhineland High Commission. When Major General Allen became American Representative in June, 1920, he appointed Colonel Stone his deputy and representative on the High Commission. In addition to his duties with the Rhineland Commission, Colonel Stone, in September 1920, also was appointed Officer in Charge of Civil Affairs, a position which he still holds.

The Belgian Government expressed its appreciation of Colonel Stone's work with the High Commission by making him an officer of the Order of Leopold.

MR. MANTON DAVIS.

Manton Davis, son of Robert Thomas and Sarah Elizabeth Davis, was born at Mayfield, Kentucky, 15 July 1876. His academic education was received in private schools and at West Kentucky College. He studied law at University of Virginia, graduating June 1901 with degree of Bachelor of Laws. He engaged in the general practice of law at St. Louis, Missouri, from 1901 until the entry of the United States into the World War; entered the first officers' training camp at Ft. Riley, Kansas, May 1917; commissioned Captain of Infantry August 1917 and assigned to 89th Division then in process of formation at Camp Funston, Kansas; continued with this division throughout its training, came with it to France and remained with it until after the Armistice and the occupation of Germany; commanded at various times Company A, Machine Gun Company, Headquarters Company, Second Battalion, all of the 354th Infantry; participated in St. Mihiel Defensive, St Mihiel Offensive and Meuse-Argonne Offensive. On the occupation of Germany in December, 1918, he was relieved from duty with troops and was made Officer in Charge of Civil Affairs, 177th Infantry Brigade, Priim, Germany; was transferred April, 1919, to Headquarters, 3rd Army, and made Officer in Charge of Civil Affairs, Coblenz; promoted to Major, Infantry, May 1919; September 1919, was appointed Legal Adviser to Officer in Charge of Civil Affairs, American Forces in Germany.

In October, 1919, preparation was begun for the taking over the control of the Occupied Territories by the Interallied

Rhineland High Commission provided for in the Rhineland Agreement annexed to the Treaty of Peace. The principal work of preparation was the drafting of ordinance to become operative when the treaty of peace should enter into effect. The services of Major Davis were loaned by the American Army to the Rhineland Commission to assist in this work. He was demobilized 1st November, 1919, and appointed American Legal Adviser to the Interallied Rhineland High Commission. In June, 1920, General Allen was made American Representative, I.A.R.N.C., Mr. Noyes and Mr. Day, respectively, Rhineland Commissioner and Deputy Commissioner returning to the United States. When this change was effected in addition to his duties as Legal Adviser, Mr. Davis was made Executive Officer of the American Department, and has continued so to serve.

MR. PIERREPONT B. NOYES.

Mr. Pierrepont B. Noyes was born on August 18, 1870 at Oneida, N. Y. He graduated from Colgate University in the class of 1892. In 1896 he became General Manager of the Oneida Community and in 1910, became president of the concern. In 1909, he was President of the American Hardware Manufacturers' Association.

During the war he was assistant to the Coal Administrator, H. A. Garfield, and Director of Conservation in the Fuel Administration from August, 1917 to December. 1918. He was a member of the Industrial Priority Board of the War Industries Board during 1918. As a member of the Joint War Labor Board, he was appointed to draft for President Wilson the national war labor policy.

He was nominated by the Democratic Convention of 1918 for Congress from a New York constituency, but declined to run. While the peace negotiations were still in progress, he was appointed American Delegate to the Interallied Rhineland Commission, on April 28, 1919, and held that position until his return to the United States in June, 1920.

MR. WALLACE H. DAY.

Mr. Wallace H. Day was born on October 4, 1891 at Leesville, Connecticut. Previous to the war, he engaged in business in New York City as Manager and Director of Day & Meyer, Inc.

During the war, Mr. Day was Chief of Customs and Postal regulations of the United States War Trade Board, from November, 1917 to January, 1918, at which time he was appointed Assistant Director of the Bureau of Customs and Trade Adviser to the Bureau of Exports of the War Trade Board, in charge of regulations pertaining to railroads, steamship lines, customs service, postal control and shipments in transit.

In February, 1919, Mr. Day was appointed a member of the American Commission to Negotiate Peace, as a special assistant to Mr. Vance C. McCormick. In March, 1919, he was appointed American Representative on the Interallied Economic Committee for Occupied Territory at Luxemburg and representative of the United States War Trade Board in the Occupied Territories of the Left Bank of the Rhine.

On May 10, 1919, he was appointed Deputy American Commissioner on the Interallied Rhineland Commission. He held this office until his return to the United States in June, 1920.

BARON ROLIN-JAEQUEMYNS.

Baron Rolin Jaequemyns, Belgian High Commissioner in the Occupied Territories, was born at Ghent in 1863. His father was Minister to His Majesty, King Leopold II, and his grandfather, Minister to His Majesty, King Leopold I.

Monsieur Rolin Jaequemyns attended school in Ghent and Paris (Lycee Charlemagne) and then studied natural sciences and law at Brussels University.

Doctor of Laws and Counsellor to the Cour d'Appel (Court of Appeal) in Brussels, he made a specialty of international law, in which branch his father had built up a great reputation for himself.

His inclinations led him to study foreign political questions, and in 1889 he was one of the delegates to the first Peace Conference at the Hague, where he played a prominent role as examiner and reporter of the rules of land and naval warfare.

For more than twenty years, he edited the "Revue de Droit International" (International Law Review), the publication of which was interrupted for the entire duration of the German occupation of Belgium; he was successively elected member of the Institut de Droit International (Institute of International Law), of the Commission for Diplomatic Inquiry attached to the Belgian Ministry for Foreign Affairs, of the Superior Council of the Congo, and of the Colonial Council.

At the very beginning of the war, inspired by the precepts of the rules of land and naval warfare to which he had contributed, under the auspices and with the active assistance of the Countess Jean de Merode, wife of the Lord Marshal of the Belgian Court, he founded the Belgian Agency of Information and Relief for Prisoners of War, which, through the war, was the intermediary approved by the Belgian Government and authorized by the Germans between Belgian private relief and all the Belgian and Allied prisoners of war who had their homes in Belgium. The donations sent by this society amounted to some 50,000,000 francs and many French and English people were thus assisted.

During the Peace Conference, Baron Rolin Jaequemyns held the office of Secretary-General to the Belgian Delegation, then that of Plenipotentiary.

Baron Rolin Jaequemyns is a Commander of the Legion of Honor and of the Russian Order of St. Anne; and a Grand Officer of the Order of the Sacred Treasure of Japan. In addition, he is a Commander of the Order of Leopold and is authorized to add the gold bar to this decoration because of services rendered during the war.

MONSIEUR FERNAND CATTOIR.

Monsieur Fernand Cattoir was born in Brussels on December 4, 1873. Doctor of Laws and Counseller to the Cour d'Appel at Brussels, he specialized in the study of public and administrative law. In 1910, he became the Chef de Cabinet of the Minister for Home Affairs, Monsieur Paul Berryer.

During the war, he accompanied the Belgian Government to Antwerp and to France, where he directed the Administrative Section of the Ministry for Home Affairs and participated in the drawing up of the law decrees of Le Havre. He then became Chef de Cabinet of the Count de Broqueville, Minister for Home Affairs after the Armistice.

In the month of June, 1919, he was made "Advocate Counsellor" (Avocat-conseil) to the Minister for Home Affairs and to the Bureau for the Devastated Regions.

At present he is secretary to the "Carnegie Hero Fund" for Belgium.

For several years, he was Professor of Civil Law at the Advanced School of Commerce of the Institut St. Louis.

Since 1902, he has been a member of the Council of Mines and of the Litigations Board of the Ministries for Industry, Labor, and Agriculture.

Nominated deputy of the Belgian High Commissioner in the Rhineland Territories, Monsieur Fernand Cattoir took an active share in drawing up the first ordinances of the High Commission and helped to bring all the preliminary work to a successful finish. Recalled to take up the office of Chef de Cabinet to Monsieur Berryer, Belgian Minister for Home Affairs, in the month of January, 1922, Monsieur Cattoir was at the same time the recipient of a most unusual honor by being raised to the highest rank of the Belgian Administration, that of Director General.

Monsieur Cattoir is an Officer of the Ordre de la Couronne (Order of the Crown), an Officer of the British Empire, an Officer of the Legion of Honor, and a Chevalier de l'Ordre de Leopold (Knight of the Order of Leopold.)

COUNT RAOUL DE LIEDEKERKE.

Count Raoul de Liedekerke, Deputy of the Belgian High Commissioner, was born in Brussels on November 14, 1882.

Doctor of Laws, he was called to the Bar in Brussels and became collaborator of Monsieur Beernaert, Cabinet Minister, and of Monsieur Delacroix who has been Prime Minister since the war and who is now Belgian Delegate to the Reparation Commission.

He became a member of the Royal Cabinet.

He enlisted in the army at the time of the German invasion and took part in the 1914-1918 campaign.

He has been attached to the Interallied Rhineland High Commission since the month of February, 1920, and succeeded Monsieur Cattoir as Deputy High Commissioner on the departure of the latter in the month of January last.

SIR HAROLD ARTHUR STUART

SIR HAROLD ARTHUR STUART, K.C.M.G., K.C.S.I., K.C.V.O., born July 20th, 1860. A retired officer of the Indian Civil Service. Served as Under Secretary to the Government of India under Sir M. E. Grant-Duff and Lord Connemara; was Secretary under Lord Ampthill, and was Private Secretary to Sir Arthur Havelock. Inspector General of Police in Madras, 1898; Secretary to the Police Commission appointed by Lord Curzon in 1902; First Director of Central Criminal Intelligence Department, India, 1904; Home Secretary to the Government of India, 1908-1911. Member of the Executive Council, Madras, 1912-1916. Served in the Ministry of Food in London 1916-1918. Was appointed first

British High Commissioner on the Inter-Allied Rhineland High Commission on August 1st, 1919; resigned on October 9th, 1920. In 1921 was appointed as His Majesty's Commissioner on the Inter-Allied Plebiscite Commission, Upper Silesia.

MR. MALCOLM ARNOLD ROBERTSON.

MALCOLM ARNOLD ROBERTSON, C.M.G., born September 2, 1877. Entered Foreign Office, 1898. Was in attendance on the Representative of Honduras at the Coronation of King Edward VII, August 1902. Received the Coronation Medal. Appointed an acting Third Secretary in the Diplomatic Service, and was attached to Berlin from January 1, 1903 to December 31, 1904. Appointed Second Secretary at Peking, November 23, 1905. Transferred to Madrid, December 16, 1907, and to Bucharest, March 29, 1910, where he acted as Charge d'Affaires from September 2 to November 27, 1910, and from January 11 to October 9,1911. Transferred to Rio de Janeiro February 10, 1912. Promoted to be First Secretary, November 23, 1912. Transferred temporarily to Monte Video to take charge of the Legation, November 1, 1912; and acted as Charge d'Affaires there from November 21, 1912 to September 24, 1913. Transferred to Rio de Janeiro, October 11, 1913, where he acted as Charge d'Affaires from October 29, 1913 to April 27, 1915. Made a C. M. G. June 3, 1915. Transferred to Washington, October 14, 1915, and to The Hague, June 24, 1918. Acted as Charge d'Affaires from September 9 to October 18, from December 15, 1918 to February 15, 1919, and July 1 to October 26, 1919. Promoted to be Counsellor of Embassy in H. M. Diplomatic Service, September 13, 1919; appointed Deputy British High Commissioner on the Interallied

Rhineland High Commission, December 1, 1919, and British High Commissioner on October 10, 1920. Transferred to Tangier, December, 1921, as His Majesty's Agent and Consul-General.

LORD KILMARNOCK.

LORD KILMARNOCK, C.M.G., born October 17, 1876. Nominated an Attache, April 7, 1900. Appointed to Brussels July 15, 1901. Promoted to be Third Secretary, July 4, 1902. Acted as Charge d'Affaires at Brussels from July 7 to 8, 1903. Transferred to Vienna, July 4, 1906. Promoted to be Second Secretary, July 4, 1906. Transferred to the Foreign Office, October 1, 1907, and to Stockholm, September 15, 1909, where he acted as Charge d'Affaires for various periods from November 3, 1909 to February 1, 1913. Promoted to be a First Secretary, July 4, 1913. Transferred to Havre, November 21, 1915, where he acted as Charge d'Affaires for various periods from July 27, 1916 to November 13, 1917. Transferred to Copenhagen, July 10, 1918, where he acted as Charge d'Affaires from October 5, 1918 to March 9, 1919. Made a C.M.G. June 3, 1919. Promoted to be a Counsellor of Embassy in H.M. Diplomatic Service, September 15, 1919, and appointed H.M. Charge d'Affaires at Berlin, January 10, 1920. Transferred to Coblenz, December 1, 1921, as British High Commissioner on the Interallied Rhineland High Commission.

LIEUTENANT COLONEL RUPERT SUMMER RYAN

Lieutenant Colonel RUPERT SUMMER RYAN, D.S.O., born May 6, 1884. Entered Royal Field Artillery December, 1905 as Second

Lieutenant. Promoted Lieutenant December, 1908; Captain, September 1914; Major, September 1916; Brevet Lieut. Colonel June, 1919. Served with R.H.A. Great Britain 1906-1912; Egypt 1913-1914; France, 1914-1915; joined General Staff 1915 and served successively with 7th Division, 13th Corps, Cavalry Corps. Appointed Chief of Intelligence 1st British Army in August 1917; served in this capacity until Armistice. Appointed to Staff of Military Governor, British Occupied Territory of Germany, November 1918 and became Chief of Staff, August 1919. Appointed to Interallied Rhineland High Commission as Commissioner at Cologne January, 1920, and transferred to Coblenz as Deputy British High Commissioner, October, 1920.

MONSIEUR PAUL TIRARD.

MONSIEUR PAUL TIRARD, born in 1879, was educated at the College of Nogent-le-Rotrou, where he was a brilliant student.

As student at the School of Political Science, he came out second, a "Laureate of the School"; he performed his military service at Chartres, at the Marceau barracks; in 1902, he passed his examination for the Conseil d'Etat; he then became a Professor at the School of Political Science in Paris and was first made councillor and then Master of the Court of Claims of the Conseil d'Etat.

In 1912, General Lyautey who had been appointed Resident-General in Morocca, asked him to be his chief civilian assistant in the organization of Morocco, over which France had just assumed a protectorate. M. Tirard played a prominent part in this great work, the success of which is well known; it is based on a generous conception of the alliance of the advantages of European colonization with the respect of the rights and customs of the natives. General Lyautey conferred on him the rosette of an "Officer of the Legion of Honor".

In 1914, M. Tirard asked to be allowed to to to the Front. He was at first a lieutenant; later a captain. General Joffre entrusted the administrative organization of reconquered Alsace to him. At his own request, he was put in command of a company of _Chasseurs a pied_ (66th Battalion); he was created an Officer of the Legion of Honor for military services, obtained three citations, one of them being at Army orders, giving the right to wear a palm. Meanwhile, the Government entrusted an important economic mission to Russia to M. Tirard.

At the time of the Armistice (November, 1918) Marshal Foch entrusted M. Tirard with the general control of the administration of the territories occupied on the left bank of the Rhine by the Allied armies. After one year of this work, the government appointed M. Tirard as High Commissioner of the French Republic in the Rhineland. In this capacity, he acts as President of the Interallied High Commission.

He was made a Commander of the Legion of Honor, for military services, in July, 1921.

MONSIEUR AMEDEE ROUSSELLIER

Monsieur AMÉDÉE ROUSSELLIER. born in 1879, was educated at the Lycée Condorcet where he was a brilliant student. On leaving the Lycée, he studied law at the University of Paris where he obtained the diploma of Doctor of Political and Economic Science; at the same time he was finishing his course of study at the School of Political Science from which he came out second and a "Laureate of the School".

After having performed his military service at Nîmes, he prepared for the Conseil d'Etat, was received in 1904 and became in turn councillor and master of the Court of Claims.

In 1913, the Secretary of the Navy chose him as chief deputy in his council; he then became head of the council of the Under Secretary of State to the Ministry for War.

Mobilized on August 2, 1914, as Second Lieutenant in the Quartermaster Department, he was promoted first to the rank of Lieutenant, then to that of Captain. Detailed to a division which was fighting in Artois, he was given charge of the military administration of the town of Thann in Alsace, which had just been delivered by French troops.

As early as April, 1915, M. Roussellier, for war services, had been made a Chevalier of the Legion of Honor, with a citation at Army orders giving the right to wear a palm. The following year he was sent on a mission to Italy, from which he was recalled in 1917 to fill the office of Chef de Cabinet to the Ministry of the Colonies.

At the time of the Armistice, he was attached to M. Tirard as Deputy Controller General of the Rhineland Territories and was appointed Deputy French High Commissioner when the Interallied Rhineland High Commission was constituted on January 10, 1920.

M. Roussellier was promoted to the grade of Officer of the Legion of Honor in September, 1920.

HERR VON STARCK.

Carl von Starck, who retired last summer, (1921) was born in Cassel (Hesse-Nassaw) in 1857. After completing his studies, which took him to England, France and Switzerland, he was appointed junior official (Regierungsassessor) in the Prussian administrative service, was assigned to duty in the Landrat's office in Memel (East Prussia) and was on duty as Aide in the National Bureau of the Interior (now the National Ministry of the Interior) from 1896 to 1900. In 1901 he was appointed Landrat in Hörde (Westphalia) and in 1905 Assistant Superintendent of Police of Potsdam and later Superintendent. His appointment as Administrator (Regierungspraesident) of the Cologne District followed in 1917. Herr von Starck was able in a high degree to gain the sympathies of the population during his term of office in Cologne. In the summer of 1919 he was appointed National and Prussian State Commissioner for the Occupied Rhenish Territories. In this capacity he succeeded in mediating with great skill between the often conflicting interests.

Herr von Starck married a lady of a family of Rhenish industrials, and they have several children. At present, Herr von Starck is living on his estate located near Cassel.

Prince von Hatzfeld-Wildenburg.

The Prince of Hatzfeld-Wildenburg, a son of the well-known diplomat, the Count of Hatzfeld, whose last post held was a long assignment as German Ambassador in London, was born on June 30, 1867. The Prince served a short time as officer but speedily transferred to the diplomatic service and was attache from 1891 to 1901 at the London Embassy, and rose to the grade of Third Secretary and subsequently Second Secretary of the Embassy during his father's incumbency of the post of Ambassador. As Second Secretary he was transferred to Paris in 1902. From 1906 to 1908 he was Counsellor to the Washington

Embassy and took a conspicuous part in bringing about conclusion of the German-American Trade Agreement. He was on duty as Diplomatic Agent in Cairo from 1909 to 1912 and while there concluded the Trade Agreement with Egypt. In 1911 he married the only daughter of the former ambassador to Madrid, Baron von Stumm (of the well-known Rhenish family of industrials of that name). In 1912 he left the service and devoted himself to the management of his estates. During the war he was a Red Cross delegate to the German Governor General of Belgium and later Red Cross delegate in Sofia. In 1918, he acted as Germany's diplomatic representative at the Anglo-German negotiations concerning the exchange of prisoners and was able to bring them to a favorable close. The Prince, while sitting in the Prussian Upper House, was an adherent of the so-called Liberal faction.

In the autumn of 1921 he was appointed to succeed the retiring Herr von Starck as National and Prussian State Commissioner for the Occupied Rhenish Territories.

A P P E N D I X 11

P e r s o n n e l

of the

American Department.

Personnel of

American Department, I.A.R.H.C.,

from the time of its inception

December 31, 1921.

Name	Position	Period of Service
Allen, Henry T., Major General,	American Representative,	June 1920 to date.
Antoine, Germaine,	Intelligence,	March 1921 to date
Backman, Stanley G.,	Kreis Representative,	July 1919 to Aug. 1920.
Baker, Alvin R.,	Kreis Representative,	July 1919 to date.
Ballas, Marcelle,	Stenographer,	Mar. 1920 to June 1920.
Barton, Alfred I.,	Secretary,	July 1920 to date.
Bean, Marion O.,	Chauffeur,	Nov. 1919 to Dec. 1919.
Benton, Thomas R.,	Clerk,	May 1919 to date.
Bird, Clarence,	Intelligence,	Mar. 1921 to date.
Bissell, H. B.,	Kreis Representative,	Nov. 1920 to Nov. 1921.
Bledsoe, W.J.,	Mechanic	Sept. 1919 to Dec. 1919
Boex, Alice	Intelligence,	Mar. 1921 to date.
Bohrmann, Catherine,	Intelligence,	Mar. 1921 to date.
Borrett, Theo.,	Kreis Assistant,	July 1919 to Dec. 1920.
Buerger, August W.,	Kreis Assistant,	Nov. 1919 to Mar. 1921.
Cahill, Walter J.,	Clerk,	July 1921 to date.
Cravens, Jack,	Chauffeur,	Oct. 1920 to date.
Conrad, Michael,	Mechanic,	July 1921 to date.
Cotterell, Isabelle	Translator,	Jan. 1921 to date.
Crooks, Adrian R.,	Stenographer,	Sept. 1919 to date
Cross, John R.,	Secretary,	July 1919 to Dec. 1921.
Day, Wallace H.,	Deputy Commissioner,	May 1919 to June 1920.
Davis, Manton	Legal Adviser,	Nov. 1919 to date.
Davis, Nathaniel F.,	Kreis Representative,	July 1919 to Sept. 1920.
D'Armil, Eileen,	Clerk,	Aug. 1920 to Oct. 1920.
Dolan, John A.,	Kreis Executive,	Oct. 1919 to date.
Ealy, Harry A.,	Stenographer,	May 1920 to date.
Earny, Albert	Chauffeur	Sept. 1919 to Dec. 1919.

Name	Position	Period of Service
Easterbrook, Gladys,	Clerk,	Sept. 1919 to May 1921.
Ernst, Homer H.,	Clerk,	July 1919 to Aug. 1920.
Evans, William S.,	Office Assistant,	Sept. 1919 to Dec. 1919.
Fieker, Theodore F.,	Kreis Representative,	Dec. 1920 to date
Flint, Addison M.,	Kreis Executive,	July 1919 to June 1920.
Florin, Andrew,	Secretary,	Mar. 1920 to date.
Furlong, Mary,	Stenographer,	Oct. 1921 to date.
Fyfe, Edith M.,	Clerk,	Aug. 1919 to Nov. 1921.
Gane, Henry,	Assistant Legal Adviser,	Oct. 1919 to Dec. 1919.
Grayot, Virgil,	Chauffeur,	July 1919 to Mar. 1920.
Gebhardt, Fred W.,	Kreis Assistant,	Nov. 1919 to date.
Gilvin, Jesse C.,	Mechanic,	July 1920 to Sept. 1921.
Hansen, C. E.,	Kreis Representative,	July 1919 to Dec. 1919.
Haller, Eugene H.	Kreis Representative,	July 1919 to date.
Hampe, Adolph,	Kreis Assistant,	July 1919 to Mar. 1921.
Happel, A. P.,	Assistant Legal Adviser,	July 1919 to date.
Hartung, Albert,	Kreis Assistant,	Sept. 1919 to Mar. 1921.
Hilleglass, P.R.,	Clerk,	Sept. 1920 to Dec. 1920.
Holt, Joseph R.,	Kreis Representative	July 1919 to date.
Hostetter, Aaron T.,	Mechanic,	Oct. 1920 to date.
Hovde, Bennie O.,	Chauffeur,	July 1920 to Sept. 1921.
Ireton, Robert E.,	Assistant Legal Adviser,	Nov. 1920 to date.
Jamieson, Natt F.,	Kreis Representative,	July 1919 to Dec. 1920.
Johnson, E. M.,	Stenographer,	Oct. 1919 to Dec. 1919.
Kaczinski, Max,	Messenger,	Sept. 1919 to date
Kandel, Emil,	Messenger,	June 1921 to date.
Kauer, Otto F.,	Translater,	Sept. 1919 to Dec. 1919.
Kearney, William R.,	Kreis Representative,	Dec. 1920 to Mar. 1921.
King, Arthur R.,	Chauffeur,	Dec. 1920 to May 1921.

Name	Position	Period of Service
Kingston, Harry L.,	Chief of Administration and Finance,	Aug. 1919 to date.
Kiley, H. P.,	Stenographer,	July 1921 to date.
Knox, W. C.,	Commercial Adviser,	May 1919 to Oct. 1919.
Landt, Ernest C.,	Kreis Assistant,	Oct. 1919 to date.
Larson, Uno L.,	Intelligence,	Oct. 1919 to Dec. 1919.
Ludlum, Mrs.	Clerk,	Feb. 1920 to May 1920.
Massey, A. P.,	Intelligence,	Mar. 1921 to date.
Mattson, Hugo L.,	Chauffeur,	July 1919 to Dec. 1919.
Michel, I.C.A.,	Stenographer	June 1919 to Jan. 1920.
McLennan, Christian,	Stenographer,	June 1920 to May 1920.
Morgan, William,	Mechanic,	Feb. 1922 to date.
Murphy, Clarence J.,	Stenographer,	Oct. 1919 to Aug. 1920.
Mussaeus, William T.,	Intelligence,	July 1919 to Dec. 1919.
Nichols, Bruce S.,	Director Motor Transportation	Oct. 1921 to date.
Niels, Herman F.,	Night Watchman	Jan. 1921 to Mar. 1921.
Noyes, Howard H.,	Secretary,	Oct. 1919 to June 1920.
Noyes, Pierrepont B.,	Commissioner,	April 1919 to June 1920.
Noyes, H. T.,	Financial Adviser,	July 1919 to Oct. 1919.
Onow, Tatiana,	Translator,	Oct. 1920 to Dec. 1921.
Pierce, Irene	Stenographer,	Sept. 1919 to date
Pritchett, Julius W.,	Messenger,	July 1919 to date.
Reay, Charles R.,	Messenger,	July 1920 to Dec. 1920.
Reese, Walter,	Chauffeur,	Aug. 1920 to Sept. 1921.
Reeves, Roscoe R.,	Chauffeur,	July 1919 to Jan. 1921.
Raynes, Maitland A.,	Office Assistant,	Oct. 1921 to Feb. 1922.
Reilly, George,	Mechanic,	April 1921 to Sept. 1921.

Name	Position	Period of Service
Rennon, P. J.,	Messenger,	Dec. 1920 to Sept. 1921.
Richmond, Harriett,	Clerk,	Mar. 1921 to date.
Richmond, Mary,	Clerk,	Mar. 1921 to date.
Rivers, Frank H.,	Translator,	Oct. 1920 to date.
Robinson, Charles J.,	Chauffeur,	April 1921 to date.
Robins, J. G.,	Night Watchman,	Mar. 1921 to Feb. 1922.
Roemer, Matthias C.,	Clerk,	Oct. 1919 to date
Roche, Raymond T.,	Secretary,	July 1921 to date
Rowan, Aileen F.,	Clerk,	Aug. 1919 to Dec. 1921.
Rust, Ella,	Stenographer,	Mar. 1921 to date
Ryder, Maud,	Stenographer,	Sept. 1919 to July 1920.
Sahlman, B. A.,	Kreis Assistant,	Sept. 1919 to Mar. 1921.
Sanford, Francis T.,	Office Assistant,	Oct. 1920 to June 1921.
Sanders, Albert,	Kreis Assistant,	July 1919 to Mar. 1921.
Seebach, Oscar,	Kreis Representative	Aug. 1919 to date
Schettini, Rene C.,	Intelligence,	July 1919 to June 1920.
Siff, Harry H.,	Clerk,	Dec. 1920 to date
Small, Lawrence A.,	Office Assistant	July 1919 to Jan. 1920.
Smith, Willye A.,	Stenographer	Feb. 1920 to date
Stone, David L., Colonel,	Deputy Representative	June 1920 to date
Stroube, Esthel O.,	Kreis Assistant,	July 1919 to June 1920.
Tenetti, Virgilio,	Messenger,	Feb. 1921 to date
Turnbull, Doris R.,	Stenographer,	Dec. 1920 to Mar. 1921.
Teurneur, Marcelle,	Stenographer,	June 1919 to date
Walz, Louis G.,	Translator,	May 1921 to date.
White, Clyde W.,	Kreis Representative,	Oct. 1920 to date
Wilton, Willem B.,	Secretary,	Nov. 1919 to date
Wolsker, Frank J.,	Chauffeur,	Oct. 1919 to Dec. 1919.
Youde, Mary	Clerk,	Jan. 1920 to June 1920.
Young, Horace H.,	Messenger,	July 1919 to date
Zaiser, Carl O.,	Chauffeur & Mechanic	July 1920 to date.
Zingg, John,	Chauffeur,	July 1921 to date.

AMERICAN REPRESENTATION IN OCCUPIED GERMANY 1920-1921

Vol. I

CORRECTIONS IN FOOTNOTES TO CONFORM TO MIMEOGRAPHED EDITION:

Page 148, fn 1 AMG, IV, 280

" 149, fn 2 AMG, I, 306 and IV, 277
" " , " 3 AMG, I, ch 18, 310

" 152, fn 2 AMG, IV, App. No. 51, p. 279 (not printed)

" 153, fn 1 AMG, IV, App. No. 51, p. 279 (not printed)
" " , " 2 Ibid.
" " , " 3 Ibid.
" " , " 4 Ibid.

" 154, fn 1 AMG, I, ch 18 and IV, pp. 275 ff.

" 159, fn 1 AMG, IV, p. 275

" 161, fn 1 AMG, IV, App. No. 51, p. 279 (not printed)
" " , " 2 Ibid.

" 168, fn 1 AMG, I, ch 15, p. 256

" 176, fn 1 AMG, I, ch 10, and III, App. No. 24, p. 210

" 217, fn 1 AMG, IV, pp. 295-97

" 220, fn 1 Conference des Preliminaires de paix, Commission de la Rive Gauche du Rhin, Proces-Verbaux, p. 126
" " , " 2 Ibid., p. 139

" 222, fn 1 See above. pp. 178-80.

Made in the USA
Monee, IL
05 November 2026

67999300R00153